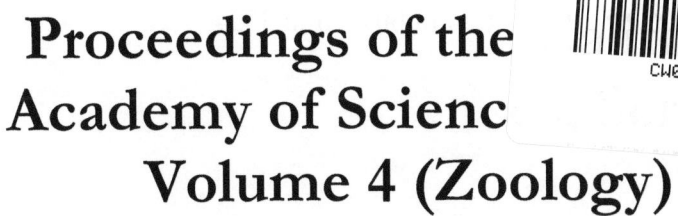

# Proceedings of the Academy of Sciences Volume 4 (Zoology)

Various

Alpha Editions

This edition published in 2024

ISBN 9789362513045

Design and Setting By
**Alpha Editions**
www.alphaedis.com
Email - info@alphaedis.com

As per information held with us this book is in Public Domain.
This book is a reproduction of an important historical work.
Alpha Editions uses the best technology to reproduce historical work
in the same manner it was first published to preserve its original nature.
Any marks or number seen are left intentionally to preserve.

# Contents

| | |
|---|---|
| INTRODUCTORY REMARKS. | - 1 - |
| SOUTH FARALLON ISLAND. | - 6 - |
| SAN MIGUEL ISLAND. | - 7 - |
| SANTA ROSA ISLAND. | - 13 - |
| SANTA CRUZ ISLAND. | - 16 - |
| ANA CAPA ISLAND. | - 17 - |
| SAN NICOLAS ISLAND. | - 18 - |
| SANTA BARBARA ISLAND. | - 18 - |
| SANTA CATALINA ISLAND. | - 18 - |
| SAN CLEMENTE ISLAND. | - 19 - |
| LOS CORONADOS. | - 20 - |
| SAN MARTIN ISLAND. | - 21 - |
| SAN BENITO ISLAND. | - 24 - |
| CERROS ISLAND. | - 26 - |
| NATIVIDAD ISLAND. | - 28 - |
| MAGDALENA ISLAND. | - 28 - |
| SANTA MARGARITA ISLAND. | - 30 - |
| SOCORRO ISLAND. | - 31 - |
| CLARION ISLAND. | - 31 - |
| PLATES IX-XI | - 54 - |

# INTRODUCTORY REMARKS.

The first contribution to the herpetology of the islands of the Pacific Coast of North America of which I have knowledge was, curiously enough, a description of the lizard of Socorro, an island perhaps the least accessible of them all. This description was published by Professor Cope in 1871. Six years later Dr. Streets recorded a few notes on the fauna of Cerros, San Martin, and Los Coronados. Since that time there have appeared at intervals contributions from Yarrow, Belding, Cope, Garman, Townsend, Stejneger, and Van Denburgh, resulting in the gradual accumulation of a considerable fund of knowledge. The papers in which this information is contained are so widely scattered through journals and the publications of various societies and museums as to be but little available. It has, therefore, been thought expedient to review the whole subject while reporting upon the material which in the last few years has been accumulating in the collection of the Academy.

In this paper there are mentioned or described twenty-nine species and subspecies, representing the fauna of eighteen islands. Of these four are amphibians, nineteen are lizards, and six are snakes.

The following forms are here described as new:—

- *Autodax lugubris farallonensis*, South Farallon Island,
- *Uta martinensis*, San Martin Island,
- *Uta stellata*, San Benito Island,
- *Sceloporus becki*, San Miguel Island,
- *Gerrhonotus scincicauda ignavus*, San Martin Island.

The island distribution of the various species and subspecies is indicated in the following table:

DISTRIBUTION OF ISLAND REPTILES AND AMPHIBIANS.

Table Key:

    A.    Farallon

    B.    San Miguel

    C.    Santa Rosa

D. Santa Cruz
E. Ana Capa
F. San Nicolas
G. Santa Barbara
H. Santa Catalina
I. San Clemente
J. Los Coronados
K. San Martin
L. San Benito
M. Cerros
N. Natividad
O. Magdalena
P. Santa Margarita
Q. Socorro
R. Clarion
S. Mainland

| Name | A | B | C | D | E | F | G | H | I |
|---|---|---|---|---|---|---|---|---|---|
| Autodax lugubris farallonensis | X | | | | | | | | |
| Batrachoseps attenuatus | | | | | | | | X | |
| Batrachoseps pacificus | | X | X | | | | | | |
| Hyla regilla | | | | X | | | | | |
| Dipsosaurus dorsalis | | | | | | | | | |
| Callisaurus ventralis | | | | | | | | | |
| Crotaphytus wislizenii | | | | | | | | | |
| Uta stansburiana | | | | X | X | | | X | X |
| Uta martinensis | | | | | | | | | |
| Uta stellata | | | | | | | | | |

| Name | J | K | L | M | N | O | P | Q | R | S |
|---|---|---|---|---|---|---|---|---|---|---|
| Uta nigricauda | | | | | | | | | | |
| Uta auriculata | | | | | | | | | | |
| Uta clarionensis | | | | | | | | | | |
| Sceloporus zosteromus | | | | | | | | | | |
| Sceloporus becki | | | X | | | | | | | |
| Sceloporus biseriatus becki | | | | X | X | | | | | |
| Phrynosoma cerroense | | | | | | | | | | |
| Gerrhonotus scincicauda | | | X | X | X | | | | | |
| Gerrhonotus scincicauda ignavus | | | | | | | | | | |
| Xantusia riversiana | | | | | | | X | X | X | X |
| Verticaria hyperythra beldingi | | | | | | | | | | |
| Cnemidophorus rubidus | | | | | | | | | | |
| Cnemidophorus multiscutatus | | | | | | | | | | |
| Cnemidophorus labialis | | | | | | | | | | |
| Bascanion anthonyi | | | | | | | | | | |
| Bascanion laterale fuliginosum | | | | | | | | | | |
| Pituophis catenifer | | | | | | | | | | |
| Crotalus exsul | | | | | | | | | | |
| Crotalus oregonus | | | | | | | | | | X | |
| Crotalus mitchellii | | | | | | | | | | |
| Autodax lugubris farallonensis | | | | | | | | | | |
| Batrachoseps attenuatus | | | | | | | | | | X |
| Batrachoseps pacificus | | | | | | | | | | ? |

| | | | | | | | | |
|---|---|---|---|---|---|---|---|---|
| Hyla regilla | | | X | | | | | X |
| Dipsosaurus dorsalis | | | | X | | | | X |
| Callisaurus ventralis | | | | | | X | | X |
| Crotaphytus wislizenii | | | | X | | | | X |
| Uta stansburiana | | | X | X | | | | X |
| Uta martinensis | X | | | | | | | |
| Uta stellata | | X | | | | | | |
| Uta nigricauda | | | | X | | | | X |
| Uta auriculata | | | | | | X | | |
| Uta clarionensis | | | | | | | X | |
| Sceloporus zosteromus | | | X | | X | X | | X |
| Sceloporus becki | | | | | | | | |
| Sceloporus biseriatus becki | | | | | | | | |
| Phrynosoma cerroense | | | X | | | | | |
| Gerrhonotus scincicauda | | | | | | | | X |
| Gerrhonotus scincicauda ignavus | ? | X | | | | | | X |
| Xantusia riversiana | | | | | | | | |
| Verticaria hyperythra beldingi | | | X | | X | | | X |
| Cnemidophorus rubidus | | | | | X | X | | X |
| Cnemidophorus multiscutatus | | | X | | | | | |
| Cnemidophorus labialis | | | X | | | | | |
| Bascanion anthonyi | | | | | | | X | |
| Bascanion laterale fuliginosum | | | | | | X | | |
| Pituophis catenifer | X | | | | | | | X |
| Crotalus exsul | | | X | | | | | |

| | | | | | | | | | |
|---|---|---|---|---|---|---|---|---|---|
| Crotalus oregonus | | | X | | | | | | X |
| Crotalus mitchellii | | | | | | | | X | X |

Little can be stated about the faunal relationships of the various islands beyond the fact that all except, probably, the Farallons are clearly Sonoran. Of the island reptiles, only fourteen are not known to live on the mainland. These are

- Autodax lugubris farallonensis,
- Batrachoseps pacificus,
- Uta martinensis,
- Uta stellata,
- Uta auriculata,
- Uta clarionensis,
- Sceloporus becki,
- Phrynosoma cerroense,
- Xantusia riversiana,
- Cnemidophorus multiscutatus,
- Cnemidophorus labialis,
- Bascanion anthonyi,
- Bascanion laterale fuliginosum,
- Crotalus exsul.

Although the evidence is thus too meager to enable one to speak positively, it would seem that the probable faunal relationship is about as follows:

TRANSITION ZONE.

*Pacific Fauna:*

Farallon Islands.

UPPER AUSTRAL ZONE.

*Californian Fauna:*

San Miguel, Santa Rosa, Santa Cruz, Ana Capa.

*San Diegan Fauna:*

San Nicolas, Santa Barbara, Santa Catalina, San Clemente.
Los Coronados, San Martin.
Perhaps San Benito, Cerros, Natividad.

LOWER AUSTRAL ZONE.

Perhaps San Benito, Cerros, Natividad.
Magdalena, Santa Margarita.
Socorro, Clarion.

## SOUTH FARALLON ISLAND.

No reptiles have been found on the Farallon Islands and it is probable that none occur there. The amphibians are represented on South Farallon Island by a salamander which has been regarded as identical with *Autodax lugubris* Hallowell. Specimens from this island, however, are much more profusely spotted or blotched with yellow than is the mainland form of this species. In examining series of specimens one finds a few individuals from the mainland as heavily spotted as some of the Farallon specimens, but the average difference seems constant and the extremes are very dissimilar. I therefore propose that the Farallon Island form be called

### 1. **Autodax lugubris farallonensis** subsp. nov.

PLATE II.

*Anaides lugubris* YARROW, Bull. U. S. Nat. Mus. no. 24, 1882, p. 158 [part].

*Autodax lugubris* COPE, Bull. U. S. Nat. Mus. no. 34, 1889, p. 185 [part]; KEELER, Zoe, v. 3, 1892, p. 154.

*Diagnosis.*—Similar to *Autodax lugubris* Hallow, but yellow spots more numerous and often larger.

*Type.*—Cal. Acad. Sci. No. 3731, South Farallon Island, Charles Fuchs, February 8, 1899.

*Description of Type.*—Head elongate, depressed, with truncate, protruding snout; nostril small, a little above and behind the corner of snout, with groove running down to edge of lip, separated from its fellow and from orbit by length of eye-slit; lip margin long and undulating; maxillary and mandibular teeth large; palatine teeth small, in series running back from each inner nostril and forming a V-shaped figure; a large well-defined patch of parasphenoid teeth divided by a slight median groove and posterior notch; tongue large, long, ovate, with a small posterior notch, free except along the median line; neck short, somewhat constricted, a well-developed gular fold; body subfusiform, diminishing toward both extremities; 13 transverse costal grooves between limbs, extending from a short distance from vertebral line entirely across belly; tail conical with similar transverse grooves; limbs well-developed, posterior longer than anterior, toes overlapping when adpressed; digits 4-5, well-developed, nearly free, with slight terminal disc-like expansion; third finger longest, first short, second and fourth nearly equal; first toe short, second and fifth and third and fourth nearly equal; skin everywhere smooth, but dotted with the mouths of small glands.

Color above smoky seal-brown, lightest on the snout and limbs, dotted, spotted and blotched with pale straw-yellow on top and sides of head, neck, body, limbs and tail; largest blotches, on sides of neck, 2 by 4 millimetres. Lower surfaces dirty yellowish white.

| | | | | | | |
|---|---|---|---|---|---|---|
| Length to anus | 72 | 38 | 58 | 66 | 67 | 75 |
| Length of tail | 64 | 33 | 50 | 52 | 56 | 71 |
| Snout to gular fold | 20 | 11 | 17 | 18 | 19 | 20 |
| Nostril to orbit | 4 | 2½ | 3 | 4 | 3½ | 4 |
| Fore limb | 21 | 13 | 18 | 19 | 20 | 22 |
| Hind limb | 24 | 15 | 20 | 20 | 22 | 24 |

Sixteen specimens were collected by Mr. Fuchs on South Farallon Island, February 8, 1899, and four by Mr. L. M. Loomis, July 9, 1896. They were found under piles of loose stone.

The spots on the type specimen are larger and somewhat more numerous than on any of the others.

SAN MIGUEL ISLAND.

I know of no records of reptiles or amphibians from San Miguel Island. Two species of lizards and a *Batrachoseps* were secured on this island by Mr. R. H. Beck while collecting for the California Academy of Sciences.

## 1. Batrachoseps pacificus *Cope*.

PLATE III.

*Hemidactylium pacificum* COPE, Proc. Acad. Nat. Sci. Phila. 1865, p. 195.

*Batrachoseps pacificus* COPE, Proc. Acad. Nat. Sci. Phila. 1869, pp. 97, 98; YARROW, Bull. U. S. Nat. Mus. no. 24, 1882, p. 153 [part?]; BOULENGER, Cat. Batrach. Grad. 1882, p. 59; COPE, Bull. U. S. Nat. Mus. no. 34, 1889, p. 129 [part?].

*Batrachoseps pacificus* was described by Professor Cope, in 1865, from a specimen said to have been collected at Santa Barbara, California. Two specimens from San Francisco were afterward referred to this species. All of the specimens I have examined from both these localities are of the common form known as *B. attenuatus*. I was, therefore, inclined to doubt the existence of *B. pacificus* as a distinct species until I examined eight specimens collected by Dr. Eisen on Santa Rosa Island in 1897. In March, 1903, Mr. R. H. Beck secured on San Miguel a large series of a *Batrachoseps* which seems to differ in no respect from that found on Santa Rosa Island, but which is very distinct from the species of the mainland.

These island salamanders agree in all important points with the original description of *B. pacificus*, but since the published descriptions of this species are not very complete, I sent a specimen from San Miguel Island to my friend Dr. Stejneger with a request that he compare it directly with the type. This he has very kindly done, and his conclusion is as follows:

"I have carefully compared it with the type of *Batrachoseps pacificus* and find them to agree completely. I have no doubt they represent the same species. As for the origin of our specimen I can only say that our record book shows the following entry: '6733. *Batrichoseps pacificus* (Type) Santa Barbara Cal. Dr. Hayes. 1881 Oct. 28. 1.' This entry is evidently made many years after the numbering of the specimen which took place in 1866, probably at the time tin-tags were substituted for the old labels most of which were destroyed as in this case. The entry is in an unknown boyish hand and is probably made from the destroyed label. The double error, *i* in *Batrachoseps* and *e* in Hays, shows that it was made by an ignoramus. I can find no other record of specimens received from the same source, but in the S. I. reports from 1864-67 I find noted that a Dr. W. W. Hays sent birds and fishes to the museum from 'Southern California'. The Santa Barbara locality is therefore not above suspicion. The other two specimens credited in Cope's Man. Batr. p. 130 to *B. pacificus*, viz. No. 4006 San Francisco, Cal. R. D. Cutts, have not been seen here since I took charge of the collection in 1889. In the record book there is entered in the remark column 'Destroyed (C)' (C) standing for Cope. The specific name *Batrachoseps pacificus* is in Cope's handwriting, while the locality San Francisco and the name of the collector are in Prof. Baird's hand."

In the light of all this it appears that the type of *Batrachoseps pacificus* may perhaps have been secured on some fishing trip from Santa Barbara to Santa Rosa or San Miguel, and that the specimens from San Francisco most probably were misidentified by Cope.

*Batrachoseps pacificus* is a larger species than *Batrachoseps attenuatus*. Its general appearance, owing to the greater broadness of head and body, is very suggestive of the various species of *Plethodon*. This resemblance is carried further in one specimen by the presence of five digits on one hind foot. Structurally, however, the species is a true *Batrachoseps*; that is to say, the tongue is adherent anteriorly, the digits are normally 4-4, the premaxillary is single, and there is a large parietal fontanelle.

> *Diagnosis.*—Costal grooves usually seventeen (rarely sixteen or eighteen); head much broader than body; color yellowish brown above, white or yellow below.
>
> *Description.*—General form elongate, slender; body cylindric or somewhat flattened; tail conical, a little longer than head and body; head depressed, rather broad, nearly circular in outline from above; snout rounded or truncate from above, truncate and high in profile; eyes large and rather prominent, separated anteriorly by about the length of the orbital slit; nostrils small, near corners of snout, separated by a little more than their distance from orbits; a very indistinct subnasal groove, not extending to margin of lip; upper jaw overhanging lower; line of lip nearly straight to below eye, then deflected downward; palatine teeth in 2 nearly straight very oblique series which nearly meet on the median line posteriorly and anteriorly do not extend to the internal nares; parasphenoid teeth separated by a narrow space posteriorly but confluent anteriorly, extending nearly to the palatine series; internal nares rather small, in front of the anterior ends of the series of palatine teeth; tongue large, oval, not emarginate, attached along the median line, free laterally and posteriorly; neck not distinct from body, with several vertical and 2 or 3

longitudinal grooves; gular fold well marked, continued forward on side of neck to eye; 1 or 2 indistinct grooves anterior to gular fold; costal grooves between limbs usually 17, occasionally 16 or 18, continued nearly to midline on back and belly; limbs short, weak, each with 4 digits; digits with rounded knob-like ends, inner digit short, rudimental, others well-developed, second and fourth nearly equal, third longest, web small or absent; tail more slender than body, with well-marked lateral grooves; a more or less indistinct dorsal longitudinal groove, most distinct on neck and pelvic region; skin smooth with minute pits; adpressed limbs widely separated.

The color above in alcoholic specimens is yellowish brown (cinnamon to mummy brown) paler on the head and limbs and often becoming fawn-color on the tail. The upper lip and all the lower surfaces are white or dull yellow. Young specimens are much darker than adults, and the lower surfaces often are minutely dotted with brown.

| | | | | | | |
|---|---|---|---|---|---|---|
| Length to anus | 25 | 36 | 49 | 52 | 52 | 56 |
| Length of tail | 20 | 31 | 64 | 56 | 63 | 59 |
| Width of head | 3 ½ | 5 | 7 | 6 ½ | 7 | 8 |
| Snout to orbit | 2 | 2 | 3 | 3 | 3 | 3 |
| Snout to gular fold | 6 | 7 ½ | 10 | 10 | 10 | 10 ½ |
| Snout to fore limb | 7 | 10 | 13 | 13 | 14 | 14 |
| Between | 15 | 22 | 31 | 36 | 33 | 38 |

| limbs | | | | | | |
|---|---|---|---|---|---|---|
| Fore limb | 5 | 7 | 9 | 9 | 8½ | 9 |
| Hind limb | 5½ | 7½ | 9½ | 9½ | 9½ | 10 |

## 2. Sceloporus becki sp. nov.

PLATE IV.

The *Sceloporus* of the mainland at Santa Barbara is the ordinary *S. occidentalis*; that is to say, it is the smaller form with a complete series of scales between the large supraoculars and the median head plates, with from thirty-five to forty-six dorsal scales between the interparietal plate and the back of the thighs, with keeled scales on the back of the thigh, and with two blue patches on the throat. Five specimens from San Miguel Island resemble this species closely in size, but are more nearly like *S. biseriatus* in coloration, and differ from both in the possession of certain characters most unusual in a member of the *S. undulatus* group. I take pleasure in naming this island form in honor of Mr. R. H. Beck, who collected the specimens.

> *Diagnosis.*—Frontal and parietal plates separated from enlarged supraoculars by a series of small scales or granules; frontoparietal plate in contact with enlarged supraoculars; scales on back of thigh smaller than those in front of anus; 43-48 dorsals between interparietal and back of thighs; scales on back of thigh keeled; whole throat and chin blue crossed by diagonal black lines which unite posteriorly with a large black patch extending across throat from shoulder to shoulder.
>
> *Type.*—Adult male, Cal. Acad. Sci. No. 4537, San Miguel Island, California, R. H. Beck, March 26, 1903.
>
> *Description.*—Head and body little depressed; nostril opening much nearer to end of snout than to orbit; upper head shields smooth, moderately large and slightly convex, interparietal largest; frontal divided transversely; parietal and frontal plates separated from enlarged supraoculars by a series of small plates or granules; frontoparietal in contact with enlarged supraoculars; superciliaries long and strongly imbricate; middle subocular very long, narrow and strongly keeled; rostral plate of moderate height but great

width; labials long, low and nearly rectangular; symphyseal large and pentangular; some series of enlarged sublabials; gulars smooth, imbricate, often emarginate posteriorly; ear-opening large, slightly oblique, with anterior denticulation of smooth acuminate scales; scales on back equal-sized, keeled, mucronate with slight denticulation, and arranged in nearly parallel longitudinal rows; lateral scales smaller and directed obliquely upward; upper and anterior surfaces of limbs with strongly keeled and mucronate scales; posterior surface of thigh with small, acuminate, keeled scales; ventral scales much smaller than dorsals, smooth, imbricate, and usually bicuspid; tail furnished with slightly irregular whorls of strongly keeled and pointed scales which are much larger and rougher above than below, where they are smooth proximally; femoral pores 14-16; 9-12 dorsal scales equaling length of shielded part of head; number of scales in a row between interparietal plate and a line connecting posterior surfaces of thighs varying from 43-48; males with enlarged postanal plates.

The color above is grayish, brownish, or greenish blue, with a series of dark brown blotches on each side of the back. A pale longitudinal band separates the dorsal from the lateral regions. The sides are brownish or grayish, mottled with darker brown and dotted or suffused with green or pale blue. The head is usually crossed by narrow brown lines, more or less irregular in distribution. A brown line connects the orbit and upper corner of the ear, and is continued backward on the neck. There is a large blue patch on each side of the belly, bordered internally with black in highly colored males. The chin and throat are blue, pale anteriorly and changing to black posteriorly, crossed by narrow oblique black lines which converge posteriorly and blend with the black patches on the throat and in front of the shoulders in males. There is a white patch at each side of the anus, and a yellowish white band along the series of femoral pores.

| | | | | |
|---|---|---|---|---|
| Length to anus | 64 | 66 | 70 | 70 |
| Length of tail | 76 | 68 | 78 | 79 |
| Snout to ear | 14 | 13 | 14 | 16 |

| | | | | |
|---|---|---|---|---|
| Width of head | 14 | 12 | 14 | 15 |
| Shielded part of head | 14 | 13 | 14 | 15 |
| Fore limb | 27 | 26 | 27 | 30 |
| Hind limb | 41 | 39 | 41 | 46 |
| Base of fifth to end of fourth toe | 16 | 15 | 16 | 18 |

This species is in general appearance similar to *S. occidentalis*, but differs in the contact of the frontoparietal and supraocular shields, the coloration of the throat, and the somewhat more feeble carination and mucronation of its dorsal and caudal scales. Specimens from Santa Rosa and Santa Cruz islands, as stated below, seem to show that this form has been developed from *S. biseriatus* stock.

Five specimens (Nos. 4534-4538) in the collection of the California Academy of Sciences were secured by Mr. R. H. Beck on San Miguel Island, March 26, 1903.

### 3. Gerrhonotus scincicauda *Skilton*.

One specimen (Cal. Acad. Sci. No. 4539) was taken by Mr. Beck on San Miguel Island, March 26, 1903. It has dorsals in 14-1/2 × 49 rows, temporals smooth, scales on arm and forearm smooth, and dark ventral lines along the middles of the scale rows. It seems to differ from the Santa Rosa Island specimens only in the slightly more feeble carination of the scales generally, the small size of the azygous prefrontal and of the scales on the under surface of the forearm, and a tendency toward the formation of fourteen rows of ventral scales shown by the presence of a few small scales along the edge of each lateral fold in addition to the usual twelve longitudinal rows. There are sixty-six ventrals in a row between the chin and the anus.

### SANTA ROSA ISLAND.

I have examined one species of *Batrachoseps* and two kinds of lizards from this island. The *Gerrhonotus* has already been reported from the island, the others are new to its known fauna.

### 1. Batrachoseps pacificus *Cope*.

Dr. Gustav Eisen secured eight specimens of *Batrachoseps* on Santa Rosa Island in June, 1897. These are now in the collection of the Academy (Nos. 3877-3880 and 3891-3894) and seem to differ in no respect from the form found on San Miguel Island. All have seventeen costal grooves.

The measurements of these specimens are

| | | | | | | | |
|---|---|---|---|---|---|---|---|
| Length to anus | 21 | 22 | 24 | 32 | 33 | 35 | 41 |
| Length of tail | 14 | 16 | 21 | 23 | 23 | ... | 46 |
| Width of head | 3 | 3 | 3 ¾ | 5 | 4 ½ | 5 | 5 |
| Snout to orbit | 1 ½ | 1 ¼ | 1 ½ | 2 | 2 ¼ | 2 | 2 ¼ |
| Snout to gular fold | 5 | 5 | 6 | 7 ¼ | 7 | 8 | 8 ½ |
| Snout to forelimb | 6 | 6 | 8 | 10 | 9 | 10 | 12 |

| | | | | | | | | |
|---|---|---|---|---|---|---|---|---|
| Between limbs | 13 | 14 | 15 | 21 | 19 | 23 | 26 | 26 |
| Fore limb | 4½ | 4 | 5 | 6¼ | 6 | 6 | 7¼ | 8 |
| Hind limb | 4½ | 4 | 5 | 6½ | 6 | 6¼ | 7½ | 8 |

## 2. Sceloporus biseriatus becki *Van Denburgh*.

A series of eight *Scelopori* collected on Santa Rosa Island by Dr. Gustav Eisen in June, 1897, seems to show that the differentiation from *S. biseriatus* has not progressed so far on this island as on San Miguel. Thus although all the adult specimens from Santa Rosa Island show the coloration of the San Miguel Island form, only two have the typical arrangement of the supraoculars, while the other six specimens have the frontoparietal separated from the enlarged supraoculars. The less highly colored young males show a single median blue throat patch, as in *S. biseriatus*, indicating that the island lizard is more closely related to that species than to *S. occidentalis*.

The fact that the characters of this form seem to be constant on San Miguel while varying toward *S. biseriatus* on Santa Rosa and Santa Cruz islands raises an interesting question in nomenclature: Should the San Miguel Island form be regarded as a species or as a subspecies? If these lizards inhabited a peninsula one would use a trinomial for them all, but since they are found on well separated islands the facts seem to be best expressed by the nomenclature adopted above.

## 3. Gerrhonotus scincicauda *Skilton*.

PLATE VII, FIGS. 3-4.

*Gerrhonotus scincicauda* VAN DENBURGH, Occas. Papers, Cal. Acad. Sci. 5, 1897, p. 106.

I am unable to distinguish six specimens (Cal. Acad. Sci. Nos. 3881-3883 and 3896-3898) collected on Santa Rosa Island from the species now known as *G. scincicauda*; that is to say, the form with fourteen longitudinal rows of scales, single interoccipital plate, large azygous prefrontal, longitudinal lines along the middle of each row of ventral scales, and smooth temporals. This clearly is the form to which Baird and Girard applied the name *G. scincicauda*, but that it is the species originally described by Skilton seems far from certain.

The specimens from Santa Rosa Island all have dorsals in fourteen longitudinal series. The number of transverse series between the interoccipital plate and the backs of the thighs is fifty in one specimen, fifty-one in three, fifty-two in one, and fifty-three in one. One has the brachial scales very weakly keeled. They were collected by Dr. Gustav Eisen in June, 1897.

### SANTA CRUZ ISLAND.

A *Hyla* and two species of lizards have heretofore been recorded as inhabiting Santa Cruz Island. Another lizard is here reported for the first time.

### 1. Hyla regilla *Baird & Girard.*

*Hyla regilla* YARROW, Bull. U. S. Nat. Mus. no. 24, 1882, p. 171; COPE, Bull. U. S. Nat. Mus. no. 34, 1889, p. 360.

Yarrow and Cope record this species as having been collected on Santa Cruz Island by Mr. H. W. Henshaw in June, 1875, but another portion of the same lot of specimens (U. S. Nat. Mus. No. 8686) is stated to be from Santa Cruz, California. Mr. Henshaw tells me he never has collected in Santa Cruz County, and that these specimens unquestionably came from Santa Cruz Island where he collected in the summer of 1875.

### 2. Uta stansburiana *Baird & Girard.*

*Uta stansburiana* YARROW, Bull. U. S. Nat. Mus. no. 24, 1882, p. 56; TOWNSEND, Proc. U. S. Nat. Mus. v. 13, 1890, p. 144; VAN DENBURGH, Occas. Papers, Cal. Acad. Sci. 5, 1897, p. 68; COPE, Report, U. S. Nat. Mus. 1898 (1900), p. 310.

The register of the United States National Museum states that two specimens of this lizard (No. 8619) were collected by Dr. O. Loew, on Santa Cruz Island in June, 1875. These lizards are still in the National collection and are of considerable interest since they, and two from Ana Capa Island, are the only ones I have seen which approach the San Benito Island *Uta* (described below) in the character of their dorsal lepidosis. That these specimens actually were collected by Dr. Loew on Santa Cruz Island is, I think, open to little doubt, since he, with Mr. H. W. Henshaw and Dr. J. T. Rothrock, visited this island in June, 1875.

A series of eight specimens collected on Santa Cruz Island, February 7, 1889, by Mr. C. H. Townsend of the U. S. Fish Commission, (U. S. Nat. Mus. Nos. 15909-15917) are all of the ordinary *Uta stansburiana* type with imbricate dorsals and mucronate caudals. Four others, taken by Mr. Joseph Grinnell at Friar's Harbor, Santa Cruz Island, are also of the usual type. These have femoral pores 13-14, 15-15, 12-13, and 15-15.

### 3. Sceloporus biseriatus becki *Van Denburgh*.

Mr. Joseph Grinnell has kindly sent me five specimens of the *Sceloporus* of Santa Cruz Island, three of which he has given to the Academy. All five show the characteristic coloration of *S. becki*. Three have the supraoculars in contact with the frontoparietals on both sides of the head, one has these scales in contact on one side but separated on the other, and the fifth specimen has granules intervening on both sides.

### 4. Gerrhonotus scincicauda *Skilton*.

*Gerrhonotus scincicaudus* YARROW, Bull. U. S. Nat. Mus. no. 24, 1882, p. 48; VAN DENBURGH, Occas. Papers, Cal. Acad. Sci. 5, 1897, p. 106.

*Gerrhonotus multicarinatus* COPE, Report, U. S. Nat. Mus. 1898 (1900), p. 525.

Yarrow and Cope record two specimens (U. S. Nat. Mus. No. 8626) collected on Santa Cruz Island by Mr. H. W. Henshaw in June, 1875. One of these is still in the National Museum, where I examined it some years ago.

## ANA CAPA ISLAND.

I believe no reptiles have been recorded from Ana Capa. Only the following species has come into my hands.

### 1. Uta stansburiana *Baird & Girard*.

Mr. Joseph Grinnell has sent me seven specimens collected on Ana Capa Island, September 4, 1903. Five of these are typical *U. stansburiana*, but the other two have dorsals similar to those of the two specimens collected by

Dr. Loew on Santa Cruz Island; that is to say, they approach in this respect the *Uta* of San Benito Island. The dorsal scales, however, are well keeled and the caudals are of the normal type. The femoral pores in the Ana Capa specimens are 14-14, 14-15, 14-?, 14-15, 14-14, 14-15, and 14-15.

### SAN NICOLAS ISLAND.

San Nicolas Island is the type locality of *Xantusia riversiana*. No other reptile has been found there.

#### 1. Xantusia riversiana *Cope.*

PLATE V, FIG. 2.

*Xantusia riversiana* COPE, Proc. Acad. Nat. Sci. Phila. 1883, p. 29; RIVERS, Am. Nat. v. 23, 1889, p. 1100; VAN DENBURGH, Proc. Cal. Acad. Sci. 2d ser. v. 5, 1895, p. 534; VAN DENBURGH, Occas. Papers, Cal. Acad. Sci. 5, 1897, p. 132; COPE, Report, U. S. Nat. Mus. 1898 (1900), p. 552.

In describing this species Cope failed to state where his specimens were collected. Rivers later assigned them to San Nicolas Island, but the matter has remained open to question. I am, therefore, very glad to be able to record the fact that Mr. Joseph Grinnell has sent me three specimens of this *Xantusia* taken by himself on San Nicolas Island, May 22-23, 1897. One of these specimens is uniform drab, with a few dark spots. The others are of the handsome striped style of coloration (see plate).

### SANTA BARBARA ISLAND.

I believe no reptiles or amphibians have been recorded from this island. I have seen only the following species:

#### 1. Xantusia riversiana *Cope.*

Mr. Joseph Grinnell has sent me four Xantusias from Santa Barbara Island. They are smaller than the specimens I have seen from the other islands, but seem to differ in no other respect. The largest is 64 mm. from snout to vent. All are dark drab above with small, discrete black spots. One shows traces of longitudinal dorsal bands near the tail.

### SANTA CATALINA ISLAND.

One salamander, two lizards, and a rattlesnake have been taken on Santa Catalina.

#### 1. Batrachoseps attenuatus *(Eschscholtz).*

A single specimen collected at Avalon, Santa Catalina Island, by Mr. A. M. Drake (Cal. Acad. Sci. No. 3726) seems indistinguishable from the mainland species. It has nineteen costal grooves, slender limbs, and narrow head. The coloration is uniform slaty brown above, paler below. Three specimens secured on this island by Mr. Fuchs differ from this one only in the slightly paler coloration.

### 2. Uta stansburiana *Baird & Girard*.

*Uta stansburiana* COPE, Report, U. S. Nat. Mus. 1898 (1900), p. 311.

This lizard has been recorded from Santa Catalina by Professor Cope. Two specimens collected at Avalon by Mr. J. I. Carlson are in the collection of the Academy (Nos. 4754 and 4755). They seem to be fairly typical *U. stansburiana* with moderately imbricate dorsals. The femoral pores are thirteen or fourteen.

### 3. Xantusia riversiana *Cope*.

*Xantusia riversiana* RIVERS, Am. Nat. v. 23, 1889, p. 1100; VAN DENBURGH, Proc. Cal. Acad. Sci. 2d ser. v. 5, 1895, p. 534; VAN DENBURGH, Occas. Papers, Cal. Acad. Sci. 5, 1897, p. 132.

I have seen no specimens of this lizard from Santa Catalina, but Mr. J. J. Rivers states that he has received several from this island.

### 4. Crotalus oregonus *Holbrook*.

*Crotalus lucifer* YARROW, Bull. U. S. Nat. Mus. no. 24, 1882, p. 76; STEJNEGER, Report, U. S. Nat. Mus. 1893 (1895), p. 447.

Yarrow records a rattlesnake as having been taken by Mr. P. Schumacher on Santa Catalina Island in 1876. Stejneger also refers to its presence there. I have seen no snakes from any of the Californian islands.

## SAN CLEMENTE ISLAND.

Two species of lizards are known from this island.

### 1. Uta stansburiana *Baird & Girard*.

*Uta stansburiana* TOWNSEND, Proc. U. S. Nat. Mus. v. 13, 1890, p. 144; VAN DENBURGH, Occas. Papers, Cal. Acad. Sci. 5, 1897, p. 68; COPE, Report, U. S. Nat. Mus. 1898 (1900), pp. 310, 311.

Two specimens were taken on San Clemente Island by C. H. Townsend in 1889. Mr. A. W. Anthony and Dr. E. A. Mearns also found the species there and sent specimens to the National Museum.

I have examined those collected by Mr. Townsend and Mr. Anthony and six specimens sent me by Mr. Joseph Grinnell, of which three are now in the collection of the Academy, and am unable to distinguish the island lizards from the form originally described by Baird and Girard. The femoral pores in three specimens are eleven, twelve, and fourteen.

## 2. Xantusia riversiana *Cope*.

PLATE V, FIG. 1.

*Xantusia riversiana* COPE, Proc. U. S. Nat. Mus. v. 12, 1889, p. 147; VAN DENBURGH, Proc. Cal. Acad. Sci. 2d ser. v. 5, 1895, p. 534; VAN DENBURGH, Occas. Papers, Cal. Acad. Sci. 5, 1897, p. 132; COPE, Report, U. S. Nat. Mus. 1898 (1900), pp. 552, 553.

This lizard was found on San Clemente by Mr. C. H. Townsend. I have examined several specimens in the collections of the University of California and the California Academy of Sciences without finding differences between them and specimens from San Nicolas and Santa Barbara Islands.

LOS CORONADOS.

I believe that only one reptile from Los Coronados is represented in collections, but I am informed that several other kinds, including *Gerrhonotus* and *Hypsiglena*, occur on these islands.

## 1. Crotalus oregonus *Holbrook*.

*Crotalus adamanteus atrox* STREETS, Bull. U. S. Nat. Mus. no. 7, 1877, p. 40; YARROW, Bull. U. S. Nat. Mus. no. 24, 1882, p. 75 [part].

*Crotalus atrox* VAN DENBURGH, Proc. Cal. Acad. Sci. 2d ser. v. 5, 1895, p. 156 [part].

*Crotalus lucifer* STEJNEGER, Report, U. S. Nat. Mus. 1893 (1895), pp. 445, 447.

*Crotalus confluentus confluentus* COPE, Report, U. S. Nat. Mus. 1898 (1900), p. 1173 [part].

Streets recorded as *Crotalus adamanteus atrox* a rattlesnake which he secured on Los Coronados. Dr. Stejneger has shown that this specimen, which is still in the National Museum, is a Pacific Rattlesnake.

## SAN MARTIN ISLAND.

The only reptile heretofore known from San Martin is a gopher snake found by Dr. Streets. The Academy has also specimens of two species of lizards from the island, both of which are here described as new. The *Uta* probably is confined to the island, while the *Gerrhonotus* seems to be found throughout the San Diegan Fauna.

### 1. Uta martinensis sp. nov.

PLATE VI.

*Diagnosis.*—Similar to *U. stansburiana* but larger; fifth toe reaching to or beyond end of second; dorsals imbricate, mucronate, strongly keeled; scales on upper surfaces of arm and thigh keeled; scales of ear-denticulation longer than the longest diameter of largest temporal; caudals large, imbricate, strongly keeled and mucronate.

*Type.*—Adult male, Cal. Acad. Sci. No. 4698, San Martin Island, Lower California, Mexico, R. H. Beck, May 3, 1903.

*Description of the Type.*—Body and head considerably depressed; snout low, rounded; nostrils large, opening upward and outward nearer to end of snout than to orbit; head plates large, smooth, nearly flat, interparietal largest; frontal divided transversely; 3 or 4 enlarged supraoculars, separated from the frontals and frontoparietals by 1 series of small plates; superciliaries long, narrow and projecting; central subocular very long, narrow and strongly keeled; rostral and supralabials long and low; 6 supralabials; symphyseal small, followed on each side by a series of 5 or more large plates which are separated from the infralabials by 1 or 2 series of sublabials; gular region covered with smooth, hexagonal or rounded scales changing to granules on the sides of the neck and to larger imbricate scales on the strong gular fold, largest on the denticulate edge of gular fold where larger than scales on belly; a group of enlarged plates in front of ear-opening; ear denticulation very long, of 3 scales, largest exceeding in length longest diameter of largest plate in front of ear; back covered centrally with nearly uniform imbricate, keeled scales which change gradually to granules on neck and sides of body, and become mucronate posteriorly; scales largest on tail, strongly imbricate, strongly keeled and mucronate above and on sides; posterior surfaces of thighs and arms covered with small granular scales similar to those on sides of body; other surfaces of limbs provided with imbricate scales, keeled on upper surfaces of arm, forearm, thigh, leg, and foot; adpressed fore limb not reaching insertion of thigh; fifth finger reaching about to end of second; fifth toe reaching to or beyond end of second; femoral pores 15; 17-23 of largest dorsals equaling shielded part of head.

Head above grayish olive; central portion of neck and back dark brown, with 2 series of rather indefinite darker brown blotches each bordered behind and sometimes laterally by pale blue scales; some scattered pale blue dots on back and upper surfaces of limbs and tail; tail marbled with brown and blue; sides mottled with brown and pale bluish yellow, forming stripes on sides of neck; chin and gular region indigo, mottled with bluish yellow at sides; postaxillary blotch blackish indigo; lower surfaces of body, limbs and tail grayish indigo.

| | |
|---|---|
| Length to anus | 62 |
| Length of tail | 92 |
| Snout to ear | 15 |
| Shielded part of head | 14 |
| Width of head | 13 |
| Fore limb | 26 |
| Hind limb | 46 |
| Base of fifth to end of fourth toe | 18 |
| Fifth toe | 10 |

Only one specimen of this *Uta* was secured.

### 2. Gerrhonotus scincicauda ignavus subsp. nov.

PLATE VII, FIGS. 1-2.

*Diagnosis.*—Similar to *G. scincicauda* but with scales generally more strongly carinate; temporal scales keeled; dorsal [Pg 20] and caudal scales strongly keeled; scales of arm and forearm keeled; lower lateral caudals keeled; dorsals in 14 (sometimes 12-2/2) longitudinal rows; dark lines along the middles of ventral rows; azygous prefrontal large; interoccipital single; back usually with complete dark cross-bands.

*Type.*—Cal. Acad. Sci. No. 4699, San Martin Island, Lower California, Mexico, R. H. Beck, May 3, 1903.

*Description.*—Body long and rather slender, with short limbs and very long tail; head pointed with flat top and nearly vertical sides, its temporal regions often greatly swollen in old specimens; rostral plate rounded in upper outline; on top of head behind rostral a pair of small internasals, a pair of small frontonasals, a very large azygous prefrontal, a pair of large prefrontals, a long frontal, a pair of frontoparietals, 2 parietals separated by an interparietal, a pair of occipitals, and an interoccipital; 2 series (of 5 and 3) supraoculars and a series of small superciliaries; temporal scales keeled, lower sometimes only weakly; upper labials much larger than lower; 2 series of large sublabial plates below infralabials, lower larger; gular scales smooth and imbricate; scales on arm and forearm keeled; scales on upper surfaces and sides of neck, body and tail large, rhomboidal, slightly oblique, strongly keeled, strengthened with bony plates, and arranged in both transverse and longitudinal series; number of longitudinal dorsal series $12^2/_2$-14; number of transverse series between interoccipital plate and backs of thighs 42-43; a band of granules along each side from large ear-opening to anus, usually hidden by a strong fold; ventral plates about size of dorsals, smooth, imbricate and arranged in 12 longitudinal series; number

The ground color above is olive-brown, more grayish on the sides, crossed by from 9-11 dark bands. These dark bands may be brown or brownish black, continuous or broken, and are darker laterally, where their scales are tipped with white. Tail proximally marked like back, distally unicolor. Head and limbs unicolor or with traces of olive-brown mottlings. Lower surfaces suffused with gray, edges of scales lighter, darker gray or slate-colored lines along the middle of each longitudinal scale row.

| | | | |
|---|---|---|---|
| Length to anus | 103 | 110 | 117 |
| Length of tail | 167 | 128 | 125[9a] |

| | | | |
|---|---|---|---|
| Snout to ear | 21 | 25 | 26 |
| Width of head | 14 | 19 | 20 |
| Head to interoccipital | 17 | 20 | 21 |
| Fore limb | 27 | 30 | 33 |
| Hind limb | 34 | 38 | 41 |
| Base of fifth to end of fourth toe | 11 | 12 | 13 |

The three specimens of *Gerrhonotus* from San Martin Island are very similar to the species now known as *G. scincicauda*, but are much rougher than specimens from central and northern California. Reëxamination of the Californian material at hand shows that the San Martin Island form is found throughout the San Diegan Fauna and the western slope of the southern Sierra Nevada below the range of *G. palmeri*. It may be distinguished from its more northern relative by the following synopsis of characters:—

> a.—Temporals smooth; scales on arm smooth; scales on forearm smooth or weakly keeled; lateral caudals five scales behind anus smooth 6-9 rows from inferior mid-caudal line.
>
> **G. scincicauda.**
>
> a.²—Temporals keeled; scales on arm keeled; scales on forearm keeled; lateral caudals 5 scales behind anus smooth only 4-5 rows from inferior mid-caudal line.
>
> **G. s. ignavus.**
>
> ### 3. Pituophis catenifer (*Blainville*).

*Pituophis sayi bellona* STREETS, Bull. U. S. Nat. Mus. no. 7, 1877, p. 40; YARROW, Bull. U. S. Nat. Mus. no. 24, 1882, p. 106; COPE, Report, U. S. Nat. Mus. 1898 (1900), p. 876.

*Pituophis catenifer deserticola* VAN DENBURGH, Proc. Cal. Acad. Sci. 2d ser. v. 5, 1895, P. 149.

A young gopher snake taken on San Martin Island by Dr. Streets is still in the National Museum. The Academy has an adult specimen (No. 4702) collected there by Mr. Beck, May 3, 1903.

SAN BENITO ISLAND.

I know of no records of reptiles from San Benito. The Academy has received specimens of but one kind of lizard, which is here described as new.

### 1. Uta stellata sp. nov.

PLATE VIII.

*Diagnosis.*—Similar to *U. stansburiana*, but with dorsal scales not imbricate, not mucronate, often separated by minute granules, a few of the dorsal rows weakly keeled; caudals weakly keeled and very shortly mucronate, not imbricate; fifth toe not reaching end of second.

*Type.*—Adult male, Cal. Acad. Sci. No. 4704, San Benito Island, Lower California, Mexico, R. H. Beck, May 6, 1903.

*Description.*—Body and head considerably depressed; snout low, rounded and rather long; nostrils large, opening upward and outward nearer to end of snout than to orbit; head plates large, smooth, nearly flat, interparietal largest; frontal divided transversely; 4 or 5 enlarged supraoculars, separated from the frontals by 1 and from the frontoparietals by 2 series of granules; superciliaries long, narrow and projecting; central subocular very long, narrow and strongly keeled; rostral and supralabials long and low; 6 or 7 supralabials; symphyseal moderately small, followed by 2 or 3 pairs of larger plates separated from the infralabials by 1 or 2 series of moderately enlarged sublabials; gular region covered with small, smooth, hexagonal or rounded scales which change gradually to granules on sides of neck and to larger imbricate scales on the strong gular fold, largest on denticulate edge of gular fold where somewhat larger than ventrals; several enlarged plates in front of ear-opening; ear denticulation short, of 3 scales, the largest not exceeding in length diameter of largest plate in front of ear; back covered with tubercular scales of nearly uniform size becoming granular toward neck and sides of body, scales of central rows very weakly keeled, not imbricate, not mucronate, often separated by minute granules; scales largest on tail, very weakly keeled, shortly mucronate above and on sides, not imbricate; posterior surfaces of thighs and arms covered with small granular scales similar to those on sides of body; other surfaces of limbs provided with imbricate scales, smooth on arm and nearly smooth on forearm and thigh, keeled on upper surface of leg; femoral pores 15 and 16; 26-30 largest dorsals equal shielded part of head; fifth finger not reaching end of second; fifth toe not reaching end of second; adpressed fore limb not reaching insertion of thigh.

Head above uniform olive-brown; central portion of the neck, back and base of tail with a uniform brown ground with thickly scattered dots of pale blue on single scales; sides yellowish brown with scattered scales of pale yellow; upper surfaces of limbs and tail light brown dotted with pale blue; chin and gular region deep indigo with yellowish marks laterally and on labials; lower surfaces of body and limbs grayish indigo; large postaxillary blotch of blackish indigo.

*Female.*—Similar in all respects except femoral pores 13 and 15; light dots on back, limbs and tail indistinct; 2 rows of dark brown dorsal blotches becoming 1 row on tail; an indistinct series of brown lateral blotches; limbs with faint brown cross-bars.

| Sex | ♀ | ♂ (type) |
|---|---|---|
| Length to anus | 49 | 61 |
| Length of tail | 59 | 76 |
| Snout to ear | 11 | 14 |
| Shielded part of head | 11 | 13 |
| Width of head | 10 | 12 |
| Fore limb | 22 | 26 |
| Hind limb | 37 | 43 |
| Base of fifth to end of fourth toe | 15 | 18 |

Two specimens of this lizard were secured.

CERROS ISLAND.

One amphibian and seven reptiles have been recorded from Cerros or Cedros Island. I have no specimens from this island.

**1. Hyla regilla** *Baird & Girard.*

*Hyla regilla* STREETS, Bull. U. S. Nat. Mus. no. 7, 1877, p. 35; YARROW, Bull. U. S. Nat. Mus. no. 24 1882, p. 171; COPE, Bull. U. S. Nat. Mus. no. 34, 1889, p. 360; VAN DENBURGH, Proc. Cal. Acad. Sci. 2d ser. v. 5, 1895, p. 556.

*Hyla curla* BELDING, West Am. Scientist, v. 3, no. 24, 1887, p. 99.

Found by Dr. Streets near a spring of fresh water on the southeastern side of the island. It was also taken by Mr. Belding.

### 2. Uta stansburiana *Baird & Girard.*

*Uta stansburiana* STREETS, Bull. U. S. Nat. Mus. no. 7, 1877, p. 37; YARROW, Bull. U. S. Nat. Mus. no. 24, 1882, p. 57; BELDING, West Am. Scientist, v. 3, no. 24, 1887, p. 98; VAN DENBURGH, Proc. Cal. Acad. Sci. 2d ser. v. 5, 1895, p. 105; COPE, Report, U. S. Nat. Mus. 1898 (1900), p. 310.

This *Uta* was collected by Dr. Streets and Mr. Belding.

### 3. Sceloporus zosteromus *Cope.*

*Sceloporus clarki clarki* BELDING, West Am. Scientist, v. 3, no. 24, 1887, p. 99.

*Sceloporus zosteromus* VAN DENBURGH, Proc. Cal. Acad. Sci. 2d ser. v. 5, 1895, p. 110; BOULENGER, Proc. Zool. Soc. Lond. 1897, p. 498; MOCQUARD, Nouv. Arch. Mus. sér. 4, v. 1, 1899, p. 314.

This lizard has been taken only by Mr. Belding.

### 4. Phrynosoma cerroense *Stejneger.*

*Phrynosoma hernandezi* BELDING, West Am. Scientist, v. 3, 1887, p. 99.

*Phrynosoma cerroense* STEJNEGER, N. Am. Fauna, no. 7, 1893, p. 187; VAN DENBURGH, Proc. Cal. Acad. Sci. 2d ser. v. 5, 1895, p. 119; COPE, Report, U. S. Nat. Mus. 1898 (1900), p. 428, fig. 75.

This form is known from a single specimen collected by Mr. Belding.

### 5. Verticaria hyperythra beldingi *(Stejneger).*

*Verticaria beldingi* STEJNEGER Proc. U. S. Nat. Mus. 1894, p. 17.

*Verticaria hyperythra beldingi* VAN DENBURGH, Proc. Cal. Acad. Sci. 2d ser. v. 5, 1895, p. 131.

Cerros Island is the type locality of this form.

### 6. Cnemidophorus multiscutatus *(Cope)*.

*Cnemidophorus tessellatus multiscutatus* COPE, Trans. Am. Philos. Soc. v. 17, art. 3, 1892, p. 38; COPE, Report, U. S. Nat. Mus. 1898 (1900), p. 586.

*Cnemidophorus multiscutatus* VAN DENBURGH, Proc. Cal. Acad. Sci. 2d ser. v. 5, 1895, p. 126.

Professor Cope described this form from specimens secured on Cerros Island.

### 7. Cnemidophorus labialis *Stejneger*.

*Cnemidophorus labialis* STEJNEGER, Proc. U. S. Nat. Mus. 1889, p. 643; COPE, Trans. Am. Philos. Soc. v. 17, art. 3, 1892, p. 51; VAN DENBURGH, Proc. Cal. Acad. Sci. 2d ser. v. 5, 1895, p. 128; COPE, Report, U. S. Nat. Mus. 1898 (1900), p. 610.

Cerros Island is the type locality of this species also. Five specimens were collected by Mr. Belding.

### 8. Crotalus exsul *Garman*.

*Crotalus exsul* GARMAN, Mem. Mus. Compar. Zool. Camb. v. 8, no. 3, 1883, pp. 114, 174; GARMAN, Bull. Essex Inst. v. 16, no. 1, 1884, p. 35; VAN DENBURGH, Proc. Cal. Acad. Sci. 2d ser. v. 5, 1895, p. 157.

Under this name Garman has described from two specimens a small rattlesnake from Cerros Island. It seems very closely related to *C. atrox*.

## NATIVIDAD ISLAND.

I have seen only one lizard from this island.

### 1. Uta stansburiana *Baird & Girard*.

A single specimen (Cal. Acad. Sci. No. 4705) of this *Uta* was secured on Natividad by Mr. R. H. Beck, May 9, 1903.

## MAGDALENA ISLAND.

I have elsewhere recorded six species of lizards from this island. It is necessary only to mention them here. The specimens are in the collection of the Academy.

### 1. Dipsosaurus dorsalis *Baird & Girard.*

*Dipsosaurus dorsalis* VAN DENBURGH, Proc. Cal. Acad. Sci. 2d ser. v. 5, 1895, p. 93.

One was secured by Mr. Bryant in March, 1889.

### 2. Crotaphytus wislizenii *Baird & Girard.*

*Crotaphytus copeii?* VAN DENBURGH, Proc. Cal. Acad. Sci. 2d ser. v. 5, 1895, p. 95.

Upon reëxamination, I am unable to separate two specimens from Magdalena Island from the common form of this lizard.

### 3. Uta nigricauda *Cope.*

*Uta nigricauda* VAN DENBURGH, Proc. Cal. Acad. Sci. 2d ser. v. 5, 1895, p. 108.

Mr. Bryant secured a number of these lizards on Magdalena Island in 1888 and 1889.

### 4. Sceloporus zosteromus *Cope.*

*Sceloporus zosteromus* VAN DENBURGH, Proc. Cal. Acad. Sci. 2d ser. v. 5, 1895, p. 110; BOULENGER, Proc. Zool. Soc. Lond. 1897, p. 499; MOCQUARD, Nouv. Arch. Mus. sér. 4, 1899, P. 314; COPE, Report, U. S. Nat. Mus. 1898 (1900), p. 358.

The Academy has eight examples of this lizard taken on Magdalena by Mr. Bryant in February and March, 1889.

### 5. Verticaria hyperythra beldingi *(Stejneger).*

*Verticaria hyperythra beldingi* VAN DENBURGH, Proc. Cal. Acad. Sci. 2d ser. v. 5, 1895, p. 132.

Three specimens were collected by Mr. Bryant in March, 1889.

### 6. Cnemidophorus rubidus *(Cope).*

*Cnemidophorus rubidus* VAN DENBURGH, Proc. Cal. Acad. Sci. 2d ser. v. 5, 1895, p. 127.

A lizard of this species was taken on Magdalena Island in March, 1889, by Mr. W. E. Bryant.

### Santa Margarita Island.

Five reptiles are known from this island. I have not seen specimens of the *Bascanion* and cannot judge of its distinctness.

#### 1. Callisaurus ventralis (*Hallowell*).

*Callisaurus ventralis* VAN DENBURGH, Proc. Cal. Acad. Sci. 2d ser. v. 5, 1895, p. 98.

A female of this species, taken on Santa Margarita by Mr. Bryant, March 5, 1889, is in the collection of the Academy.

#### 2. Sceloporus zosteromus *Cope*.

*Sceloporus zosteromus* VAN DENBURGH, Proc. Cal. Acad. Sci. 2d ser. v. 5, 1895, p. 110; BOULENGER, Proc. Zool. Soc. Lond. 1897, p. 499; MOCQUARD, Nouv. Arch. Mus. sér. 4, 1899, p. 314; COPE, Report, U. S. Nat. Mus. 1898 (1900), p. 358.

Two examples were secured by Mr. Bryant on Santa Margarita, March 1, 1889.

#### 3. Cnemidophorus rubidus *Cope*.

*Cnemidophorus tessellatus rubidus* COPE, Trans. Am. Philos. Soc. 1892, p. 36, pl. XII, fig. F; COPE, Report, U. S. Nat. Mus. 1898 (1900), p. 584, fig. 110.

*Cnemidophorus rubidus* VAN DENBURGH, Proc. Cal. Acad. Sci. 2d ser. v. 5, 1895, p. 127.

Santa Margarita Island is the type locality of this species, which was described from seven specimens brought back by the *Albatross*.

#### 4. Bascanion laterale fuliginosum (*Cope*).

*Bascanion laterale* COPE, Proc. U. S. Nat. Mus. v. 12, 1889, p. 147.

*Zamenis lateralis fuliginosus* COPE, Am. Nat. v. 29, 1895, p. 679; COPE, Report, U. S. Nat. Mus. 1898 (1900), p. 809, fig. 178.

This snake was described from two specimens taken by the naturalists of the *Albatross*. I have seen none.

### 5. Crotalus mitchellii *Cope.*

*Crotalus mitchellii* VAN DENBURGH, Proc. Cal. Acad. Sci. 2d ser. v. 5, 1895, p. 160; COPE, Report, U. S. Nat. Mus. 1898 (1900), p. 1196.

A single rattlesnake of this species, taken by Mr. W. E. Bryant in February, 1889, is the only record for this island.

### SOCORRO ISLAND.

The following lizard is the only reptile known from this island.

### 1. Uta auriculata *Cope.*

*Uta auriculata* COPE, Proc. Bost. Soc. Nat. Hist. v. 14, 1871, p. 303; BOULENGER, Cat. Liz. Brit. Mus. v. 2, 1885, p. 214; COPE, Bull. U. S. Nat. Mus. no. 32, 1887, p. 35; TOWNSEND, Proc. U. S. Nat. Mus. v. 13, 1890, p. 143; COPE, Report, U. S. Nat. Mus. 1898 (1900), p. 300.

This *Uta* was first described by Cope in 1871 from material collected by Grayson. Townsend secured nine specimens which are now in the National Museum. The California Academy of Sciences has seventeen, taken by its expedition to the Revilla Gigedo Islands in 1903.

### CLARION ISLAND.

Although smaller than Socorro and farther from the mainland, Clarion Island is better supplied with reptiles than its larger neighbor, since it possesses a snake as well as a *Uta*, while Socorro has only a *Uta*.

### 1. Uta clarionensis *Townsend.*

*Uta clarionensis* TOWNSEND, Proc. U. S. Nat. Mus. v. 13, 1890, p. 143; STEJNEGER, Proc. U. S. Nat. Mus. v. 23, 1901, p. 715.

This lizard was first collected by Mr. C. H. Townsend who described it from five specimens. Mr. A. W. Anthony also secured it, in 1897, and sent specimens to the National Museum. The Academy has three taken by Mr. Beck.

### 2. Bascanion anthonyi *Stejneger.*

*Bascanion anthonyi* STEJNEGER, Proc. U. S. Nat. Mus. v. 23, 1901, p. 715.

Dr. Stejneger described this snake from thirteen specimens sent to the National Museum by Mr. Anthony. The Academy has eight examples of the species. The scale-rows are seventeen in all these specimens, while the gastrosteges vary from one hundred and eighty-seven to one hundred and ninety-six and the urosteges from ninety-three to one hundred and seven.

## EXPLANATION OF PLATE II.

*Autodax lugubris farallonensis* subsp. nov.

*Type*, Cal. Acad. Sci. no. 3731, South Farallon Island, California, Charles Fuchs, February 8, 1899.

Fig. 1. General view, natural size.

Fig. 2. Head from above, × 2.

Fig. 3. Head from below, × 2.

Fig. 4. Head from side, × 2.

Fig. 5. Mouth, × 2-1/2.

Fig. 6. Hind limb, × 2.

Fig. 7. Fore limb, × 2.

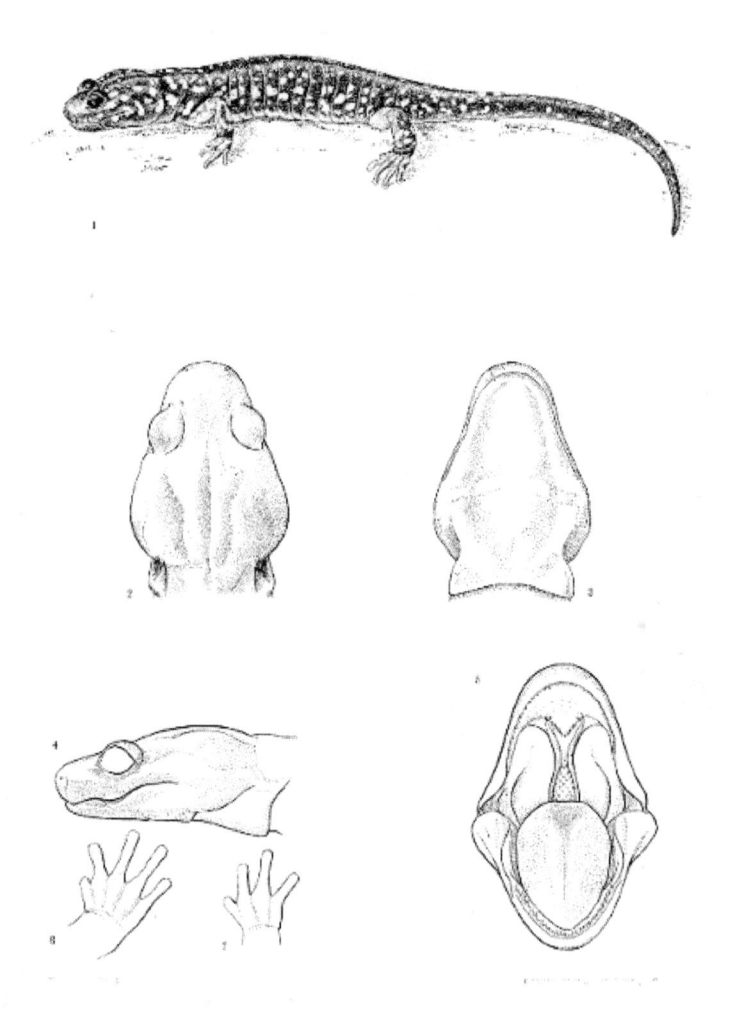

MARY WELLMAN. DEL
REY, S.F.

PHOTO.-LITH. BRITTON &

## EXPLANATION OF PLATE III.

*Batrachoseps pacificus* Cope.

Cal. Acad. Sci. no. 4601, San Miguel Island, California, R. H. Beck, March 23, 1903.

Fig. 1. General view, natural size.

Fig. 2. Head and neck from above, × 3.

Fig. 3. Head and neck from below, × 3.

Fig. 4. Head and neck from side, × 3.

Fig. 5. Mouth, × 3-1/2.

Fig. 6. Fore limb, × 3-1/2.

Fig. 7. Hind limb, × 3-1/2.

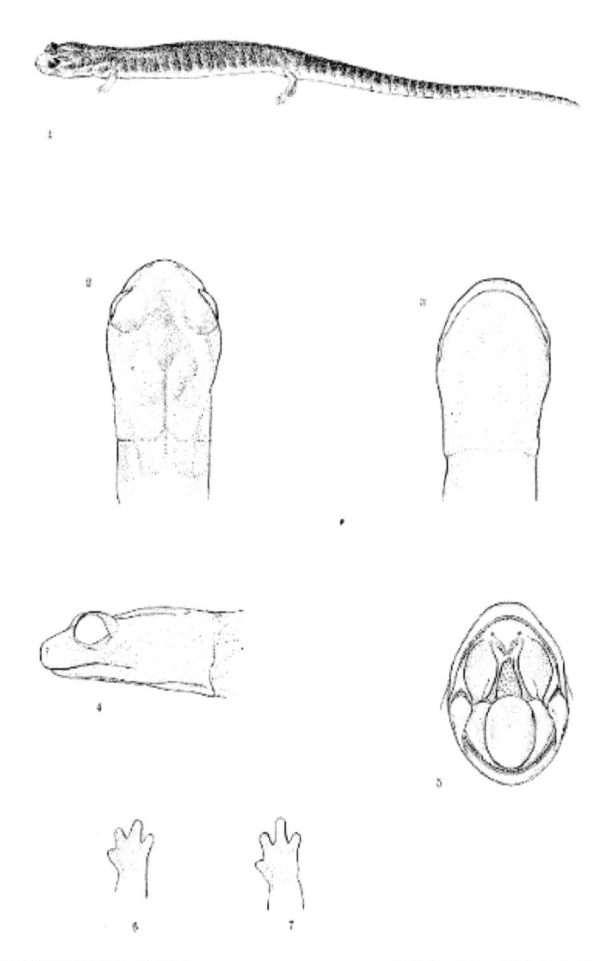

MARY WELLMAN. DEL
REY, S.F.

PHOTO.-LITH. BRITTON &

# EXPLANATION OF PLATE IV.

*Sceloporus becki* sp. nov.

*Type*, Adult male, Cal. Acad. Sci. no. 4537, San Miguel Island, California, R. H. Beck, March 26, 1903.

Fig. 1. General view, natural size.

Fig. 2. Head from below, × 2-1/2.

Fig. 3. Head from above, × 2-1/2.

Fig. 4. Head from side, × 2-1/2.

Proc. Cal. Acad. Sci. 3ᴅ. Ser. Zool. Vol. IV.

[Van Denburgh] Plate IV.

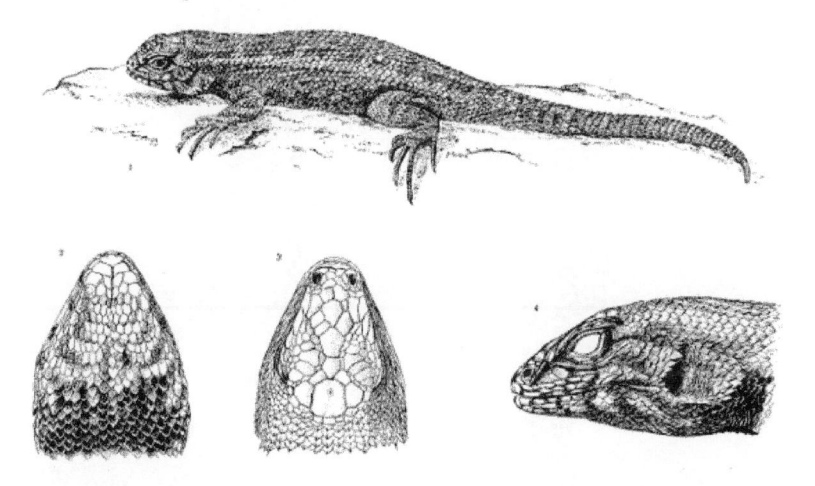

MARY WELLMAN. DEL REY, S.F.

PHOTO.-LITH. BRITTON &

# EXPLANATION OF PLATE V.

*Xantusia riversiana* Cope.

Fig. 1.  Cal. Acad. Sci. no. 3571, San Clemente Island, California. General view, × 2/3.

Fig. 2.  Cal. Acad. Sci. no. 6613, San Nicolas Island, California, Joseph Grinnell, May 22, 1897. Natural size.

PROC. CAL. ACAD. SCI. 3D. SER. ZOOL. VOL. IV.

[VAN DENBURGH] PLATE V.

MARY WELLMAN. DEL
REY, S.F.

PHOTO.-LITH. BRITTON &

## EXPLANATION OF PLATE VI.

*Uta martinensis* sp. nov.

*Type*, Cal. Acad. Sci. no. 4698, San Martin Island, Lower California, Mexico, R. H. Beck, May 3, 1903.

Fig. 1. General view, natural size.

Fig. 2. Head from above, × 2-1/2.

Fig. 3. Head from side, × 2-1/2.

Fig. 4. Scales of back, × 3.

Fig. 5. Hind limb, × 1-3/4.


PROC. CAL. ACAD. SCI. 3D. SER. ZOOL. VOL. IV.

[VAN DENBURGH] PLATE VI.

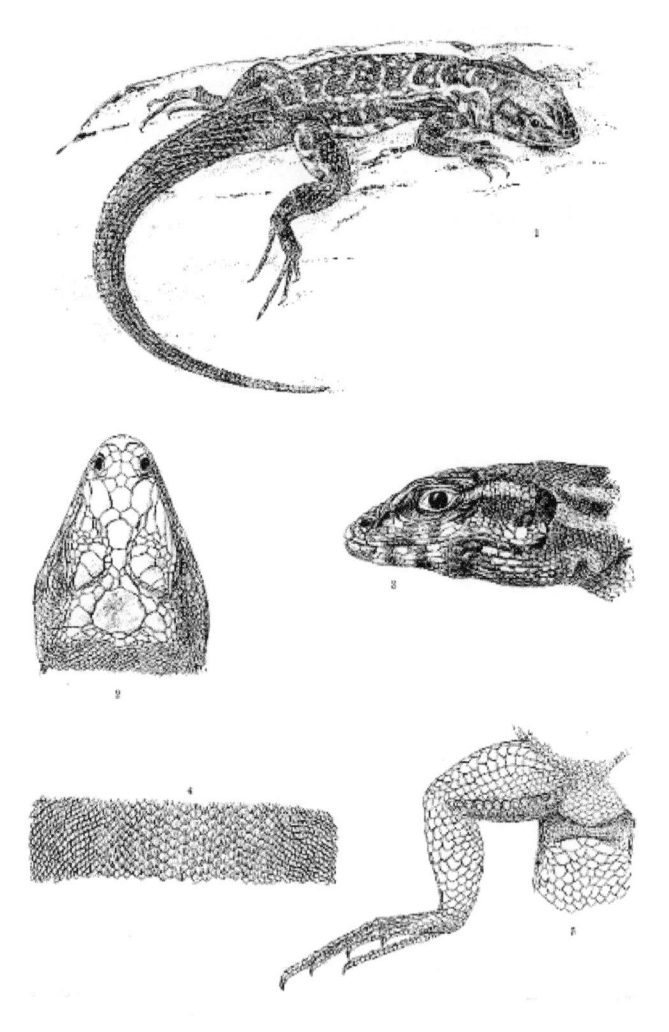

MARY WELLMAN. DEL
REY, S.F.

PHOTO.-LITH. BRITTON &

## EXPLANATION OF PLATE VII.

*Gerrhonotus scincicauda ignavus* subsp. nov.

Type, Cal. Acad. Sci. no. 4699, San Martin Island, Lower California, Mexico, R. H. Beck, May 3, 1903.

Fig. 1. Head from side, natural size.

Fig. 2. Base of tail from side, natural size.

*Gerrhonotus scincicauda (Skilton).*

Cal. Acad. Sci. no. 3897, Santa Rosa Island, California, Gustav Eisen, June 1897.

Fig. 3. Head from side, natural size.

Fig. 4. Base of tail from side, natural size.

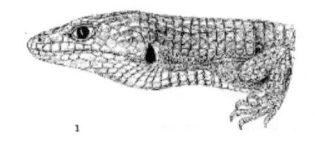

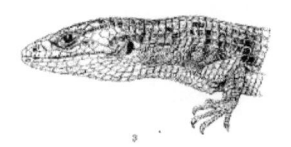

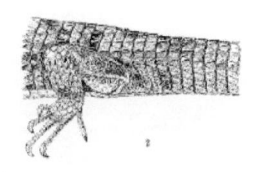

MARY WELLMAN. DEL REY, S.F.

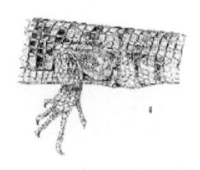

PHOTO.-LITH. BRITTON &

## EXPLANATION OF PLATE VIII.

*Uta stellata* sp. nov.

*Type*, Adult male, Cal. Acad. Sci. no. 4704, San Benito Island, Lower California, Mexico, R. H. Beck, May 6, 1903.

Fig. 1. General view, natural size.

Fig. 2. Head from side, × 3.

Fig. 3. Head from above, × 3.

Fig. 4. Scales of back, × 3.

Fig. 5. Scales of central part of back, much enlarged.

Fig. 6. Hind limb, × 1-3/4.

PROC. CAL. ACAD. SCI. 3D. SER. ZOOL. VOL. IV.

[VAN DENBURGH] PLATE VIII.

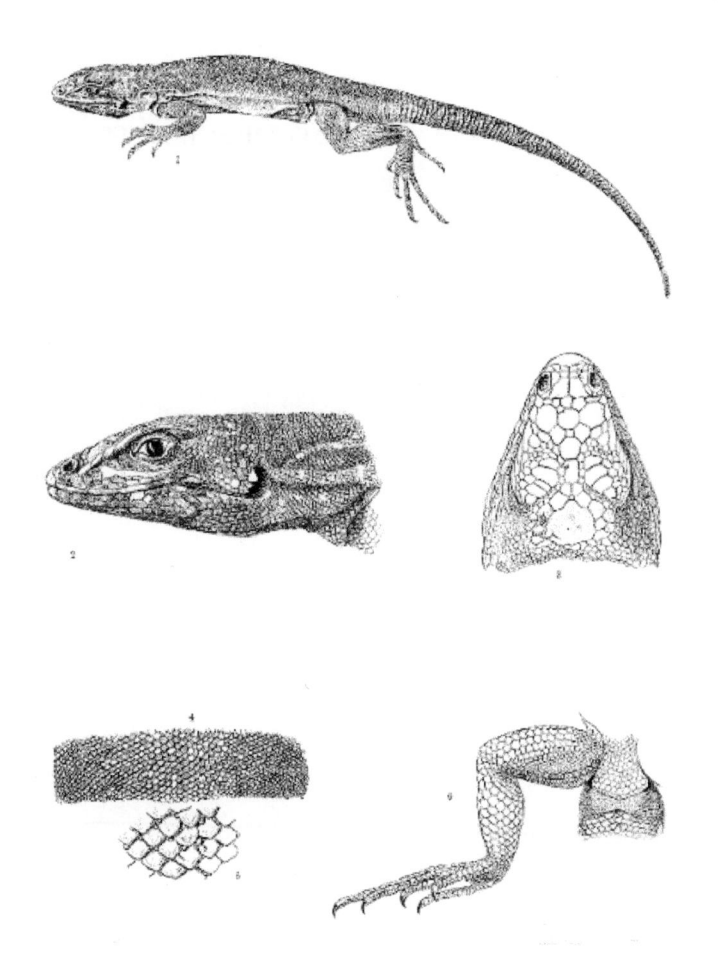

MARY WELLMAN. DEL  
REY, S.F.

PHOTO.-LITH. BRITTON &

PROCEEDINGS
OF THE
CALIFORNIA ACADEMY OF SCIENCES
THIRD SERIES

Zoology                                       Vol. IV, No. 2

*Issued December 2, 1905*

# THE SPECIES OF THE REPTILIAN GENUS ANNIELLA, WITH ESPECIAL REFERENCE TO ANNIELLA TEXANA AND TO VARIATION IN ANNIELLA NIGRA

BY JOHN VAN DENBURGH

*Curator of the Department of Herpetology.*

The genus *Anniella* was established by J. E. Gray, in 1852, to contain a single species which he named *Anniella pulchra* and described in the following terms:

"Silvery (in spirits); upper part with very narrow brown zigzag lines placed on the margin of the series of scales, the line down the center of the back and two or three on the upper part of the sides being thicker and nearly half the width of the scales.

*Hab.* California, *J. O. Goodridge, Esq., Surgeon R. N.*"

This species has since been more completely described by Bocourt, Boulenger, Cope, and Van Denburgh.

In 1885 Fischer described under the name *Anniella nigra* a specimen said to have been collected at San Diego, California. This, he stated, differed from *Anniella pulchra* in the following characters:

1. Twenty-eight longitudinal rows of scales.

2. The three median preanal scales twice as long as those preceding.

3. Tail one-third total length.

4. Color above black.

I have elsewhere stated that the number of scale rows in *Anniella pulchra* varies from twenty-four to thirty-four. The preanal scales in both the dark and light forms may be small, moderately enlarged, or twice the length of those preceding. The tail of *A. pulchra* may equal or exceed one-third of the total length of the animal. I have been unable to discover any differences in the squamation of dark and light specimens; and since the recognition of *A. nigra* as distinct from *A. pulchra* must rest solely upon the difference in pigmentation, one is tempted to inquire whether this is not merely an instance of melanism. Upon this subject I shall have more to say, but I wish first to consider certain peculiarities of squamation which have been held to distinguish another species.

*Anniella texana* was described by Mr. Boulenger, in 1887, from a single specimen labeled El Paso, Texas—a locality so far beyond the limits of the known range of the genus and of other Californian reptiles that it must be regarded with much suspicion until confirmed by the capture of additional specimens. The type of *A. texana* agrees in coloration with *Anniella pulchra*, but Mr. Boulenger finds it to differ in certain details of squamation. He assigns to it the following characters:

1. Head less depressed, snout more rounded than in *A. pulchra*.

2. A horizontal suture from nostril to second labial.

3. Frontal twice as broad as long.

4. Anterior supraocular nearly as broad as its distance from its fellow.

5. Interparietal and occipital divided (anomalously?) by a longitudinal suture.

6. Six upper labials, etc.

7. A narrow shield separates the third labial from the loreal.

8. Five lower labials.

9. Twenty-eight scales around middle of body.

10. No enlarged preanal scales.

11. Tail ending obtusely, three-eighths total length.

12. Dark gray above, with three fine black longitudinal lines; sides and lower surfaces whitish.

I will now consider these characters in connection with variations found in a series of specimens of *A. pulchra* and *A. nigra*.

1. The shape of the head and snout is subject to some variation in both *A. pulchra* and *A. nigra*. Unless the difference in shape in the type of *A. texana* is very great, one is safe in ignoring it as a basis of specific distinction.

2. One of my specimens of *A. nigra* (Cal. Acad. Sci. No. 6255) shows a horizontal suture extending from the nostril to the second labial. Another (No. 6244) has such a suture between the nostril and the rostral plate.

3. There is considerable variation in the shape and size of the frontal plate in both *A. pulchra* and *A. nigra*. It not infrequently is twice as broad as long (No. 6236, etc.), but may be nearly as long as broad. Sometimes it nearly touches the rostral (No. 5103).

4. The anterior supraocular is nearly as broad as the distance which separates it from its fellow in some specimens of *A. pulchra* (No. 5110) and *A. nigra* (Nos. 6233, 6243, 6249, etc.). In some specimens it has scarcely more than half this breadth.

5. I regard the plates which Mr. Boulenger calls interparietal and occipital as frontoparietal and interparietal, respectively. The former plate is not completely divided in any of my specimens. One example of *A. pulchra* (No. 5110), however, has it longitudinally divided throughout the posterior third of its length. The interparietal (occipital of Mr. Boulenger) is completely divided longitudinally in one example of *A. nigra* (No. 6228) and divided through one-fifth its length in another (No. 6218).

6. This is the normal arrangement, but is subject to variation.

7. This shield may be absent.

8. The number of lower labials ranges from five to seven.

9. The number of scales around the middle of the body varies in *A. pulchra* from twenty-four to thirty-four, while in 54 specimens of *A. nigra* the number is twenty-eight in 12, thirty in 36, and thirty-two in 6.

10. The preanal scales, as already stated, may be not enlarged, moderately enlarged, or twice as long as those preceding them. This is true in both *A. pulchra* and *A. nigra*.

11. The length of the tail is subject to so much variation that it cannot be regarded as furnishing a good specific character. The longest ones I have seen are one-third the total length in *A. nigra* and two-fifths in *A. pulchra*.

12. This is the coloration of some specimens of *A. pulchra*.

It will be seen that, with one exception, all of the characters of *Anniella texana* have been found in specimens of *A. pulchra* and *A. nigra* either as the normal condition or as individual variations. The single exception is the complete division of the frontoparietal plate—a condition which is manifestly anomalous, since this plate has been found partially divided in other specimens. It is evident therefore that *Anniella texana* must stand as a synonym of *A. pulchra* Gray.

*Anniella texana* being thus disposed of, one is tempted to treat *A. nigra* in the same way, regarding it as based merely upon melanistic individuals of *A. pulchra*. This view we certainly should have to adopt if both dark and light colored specimens occurred in the same localities, but I believe this has not yet been shown to be the case. Fischer, to be sure, states that the type of *A. nigra* came from San Diego, where *A. pulchra* is especially abundant, but it is quite possible that his specimen did not really originate there. All of the dark specimens I have seen, have been secured on the coast of Monterey County; and, aside from Fischer's, I know of no records of the black *Anniella* from any other locality, except Cope's reference to specimens from San Francisco. Aside from the type locality, then, it would seem that the dark form has a very limited range, being confined to the southern part of the Pacific Fauna of the Transition Zone.

In a large series of alcoholic specimens from the coast of Monterey County, I find very few showing a style of coloration similar to that of *A. pulchra*. A specimen from San Ardo, in the interior of this county, is typical of *A. pulchra*, but San Ardo is in the Upper Austral Zone. Not more than four or five of the fifty-four specimens from the coast zone could be in the least confusing, and all of these are more deeply pigmented above than is any example of *A. pulchra* before me. Forty-eight of these specimens were sent me alive, and in that condition exhibited a greater range of coloration than they show since preservation in alcohol, which seems to have intensified their dark pigmentation while dissolving the beautiful yellow of their lower surfaces. When the living lizards were received from Carmel and Point Pinos, they were divided into ten groups according to the intensity of the dorsal pigmentation, and measurements were taken of each specimen in each group. These grades of pigmentation of the living specimens, with measurements in millimeters from snout to anus and anus to tip of tail, are as follows:

1. Entire upper surface (ten, twelve, or fourteen rows of scales) and ventral surface of tip of tail very dark Indian purple. Chin and throat lighter Indian purple. More or less suffusion with Indian purple about anus. Rest of lower surfaces and sides bright gamboge yellow with chromium green staining near center of belly. Mouth flesh-color. Labials and temporals minutely dotted with iridescent greenish, silvery, or bronze. Eye black with bronze or silvery markings.

| 153 | 15 | Lateral line present | No dorsal line |

| | | | |
|---|---|---|---|
| 150 | 73 | Lateral line | Trace dorsal line |
| 150 | 70 | Lateral line | Trace dorsal line |
| 150 | 38 | Lateral line | No dorsal line |
| 147 | 40 | No lateral line | No dorsal line |
| 146 | 75 | No lateral line | No dorsal line |
| 146 | 25 | Lateral line | No dorsal line |
| 145 | 70 | Trace lateral line | No dorsal line |
| 143 | 17 | Trace lateral line | No dorsal line |
| 140 | 68 | Lateral line | No dorsal line |
| 140 | 50 | Lateral line | No dorsal line |
| 138 | 32 | Faint trace lateral lines | No dorsal line |
| 137 | 68 | Lateral line | Trace dorsal line |
| 137 | 47 | Lateral line | Trace dorsal line |
| 136 | 45 | Lateral line | No dorsal line |
| 135 | 65 | Lateral line | No dorsal line |
| 135 | 53 | Trace lateral line | No dorsal line |
| 134 | 65 | Lateral line | No dorsal line |
| 133 | 60 | Lateral line | No dorsal line |
| 132 | 63 | Lateral line | Trace dorsal line |
| 131 | 50 | Lateral line | No dorsal line |
| 131 | 16 | Lateral line | No dorsal line |
| 130 | 34 | Lateral line | No dorsal line |

2. Dark hair-brown above; bright gamboge below; chin Indian purple.

| | | | |
|---|---|---|---|
| 146 | 70 | Lateral line | No dorsal line |
| 131 | 20 | Two lateral lines | Trace dorsal line |

3. Dark purplish drab above; wax-yellow with Paris or chromium green below; chin and throat lighter Indian purple.

| | | | |
|---|---|---|---|
| 140 | 40 | Two lateral lines | Trace dorsal line |

| | | | |
|---|---|---|---|
| 134 | 20 | Lateral line | Faint trace dorsal line |
| 123 | 25 | Lateral line | No dorsal line |
| 120 | 60 | Lateral line | Fair dorsal line |

4. Hair-brown above; gamboge below; chin Indian purple.

| | | | |
|---|---|---|---|
| 130 | 65 | Two lateral lines | Dorsal line |
| 120 | 52 | Lateral line | Faint dorsal line |

5. Dark drab above; waxy gamboge below.

| | | | |
|---|---|---|---|
| 126 | 65 | Two lateral lines | Distinct dorsal line |
| 126 | 60 | Lateral line | Indistinct dorsal line |
| 125 | 20 | Two lateral lines | Faint dorsal line |
| 125 | 60 | Two lateral lines | Indistinct dorsal line |

6a. Bronzed drab above; light wax-color below; chin light Indian purple.

| | | | |
|---|---|---|---|
| 125 | 63 | Two lateral lines | Faint dorsal line |

6b. Drab above; light wax-color below; chin light Indian purple.

| | | | |
|---|---|---|---|
| 117 | 55 | Two lateral lines | Faint dorsal line |
| 116 | 55 | Two lateral lines | Faint trace dorsal line |
| 112 | 20 | Two lateral lines | Dorsal line |
| 105 | 48 | Two lateral lines | Dorsal line |

7. Grayish drab above; wax-yellow below; chin lighter Indian purple.

| | | | |
|---|---|---|---|
| 126 | 30 | Two lateral lines | Faint trace dorsal line |
| 122 | 60 | Two lateral lines | Trace dorsal line |
| 121 | 20 | Two lateral lines | Incomplete dorsal line |
| 119 | 55 | Strong lateral line | Trace dorsal line |
| 116 | 55 | Strong lateral line | Faint trace dorsal line |

8. Dark drab-gray above; dull wax-yellow below; chin light Indian purple.

| | | | |
|---|---|---|---|
| 124 | 60 | Lateral line | Faint dorsal |

line

9. Drab-gray above; straw and Naples yellow below; chin light Indian purple.

| 118 | 56 | Two lateral lines | Dorsal line |

10. Bronzed drab-gray above; pale wax-yellow below; chin light Indian purple.

| 92 | 41 | Two lateral lines | Very distinct dorsal line |

These notes show clearly that the intensity of pigmentation increases quite gradually and fairly regularly with the size of the individual, and that while young specimens may be nearly as pale as some dark individuals of *A. pulchra*, all of the large specimens are of the dark type. It is also true in a general way that the smaller the specimen the more distinctly the lines are shown.

In the light of our present knowledge, therefore, it seems necessary to regard *Anniella nigra* as a local and probably recently differentiated race rather than as a melanistic phase of *Anniella pulchra*. While the difference is purely one of color, no intergradation has yet been shown to occur in adult specimens, and the two forms must therefore be recognized as distinct species occupying separate areas in different faunal zones.

If then we ignore the localities of the type specimens of "*A. texana*" and *A. nigra*, as open to question until confirmed by the finding of additional specimens, the known distribution of the species of the genus *Anniella* is as follows:

**Anniella pulchra.**

UPPER AUSTRAL ZONE.

*San Diegan Fauna.*

San Diego County.

San Diego, Coronado, mountains near San Diego.

Riverside County.

San Jacinto.

San Bernardino County.

San Bernardino.

*Californian Fauna.*

Kern County.

Oil City to Poso Creek.

Tulare County.

Sequoia National Park.

Fresno County.

Fresno.

Monterey County (interior).

San Ardo.

San Benito County.

Bear Valley.

Contra Costa County.

**Anniella nigra.**

TRANSITION ZONE.

*Pacific Fauna.*

Monterey County (coast).

Monterey, Pacific Grove, Point Pinos, Carmel Bay.

San Francisco County.

San Francisco.

SAN FRANCISCO, CALIFORNIA,

August 18, 1905.

# PROCEEDINGS
## OF THE
## CALIFORNIA ACADEMY OF SCIENCES
### THIRD SERIES

Zoology                               Vol. IV, No. 3

*Issued December 2, 1905*

# ON THE OCCURRENCE OF THE LEATHER-BACK TURTLE, DERMOCHELYS, ON THE COAST OF CALIFORNIA

### BY JOHN VAN DENBURGH

*Curator of the Department of Herpetology.*

PLATES IX-XI

Records of the occurrence of the great marine Leather-back Turtle in the Pacific Ocean are so few that any additional observations are of much interest. Temminck and Schlegel report upon a specimen captured near the Bay of Nagasaki, Japan, in May, 1825. Mr. Swinhoe saw a large one at Amoy, China, in October, 1859. Aflalo has described a pair from Thursday Island, Queensland, Australia. Krefft mentions an example nine feet long from the coast of New South Wales. McCoy figures one caught at Portland, Victoria, Australia, in 1862. Another was harpooned by Captain Subritzky in the Bay of Islands, New Zealand, in May, 1892. Boulenger mentions a skull from the Solomon Islands. The species has been recorded from the coast of Chile by Molina and Philippi, and from Guaymas, Mexico, by Mr. Belding.

Thus it appears that the only record of the occurrence of the Leather-back Turtle in the waters of the western coast of the North American continent is the brief note by Mr. Belding in the West American Scientist, which reads as follows:

"I saw at Guaymas a Leather-back Turtle (Dermatochelys) which weighed 1,102 lbs."

I am now able to record the capture of three specimens of this turtle on the coast of California.

Early in January of the present year I received word that a large sea turtle had been caught near Santa Barbara, California, and at once arranged to purchase it for the Academy. A photograph (Plate IX) sent me at the time showed it to be a fine specimen of the Leather-back Turtle. Upon its arrival in San Francisco this turtle proved to be a female measuring six feet and seven inches from the tip of its snout to the end of its tail. Its weight was given on the bill of lading as 800 pounds, but this may have been estimated rather than actually determined. It was secured by Mr. G. W. Gourley and Albert F. Stafford, about January 2, in twenty-five fathoms of water in the open sea about two miles south of Santa Barbara.

Mr. Gourley has given me the following glowing account of its capture:

"Santa Barbara, Cal.,

"Jan. 17, 1905.

"*Dear Sir:*—Your note of 13th inst. received.... In regard to the details of the capture I will say that the turtle was first seen swimming on the surface about two miles off shore and to the southwestward of the Santa Barbara whistling buoy. I went after it (accompanied by a boy) in an 18 foot sailboat. I had a gaff with a hook on the end of it and bent about 200 feet of rope onto the handle. I had also prepared a number of other ropes with nooses on them to be ready for quick work.

"On approaching the turtle it did not hear the wash of the boat until we were within about 25 feet of it, when it made a rush to windward and started to dive, but the momentum of the boat when I luffed into the wind carried her right along side of him and I dropped the tiller and got forward with the gaff-hook and swung over the side in the weather rigging and got the hook fast in the leathery part of his neck. He immediately sounded and run out the full length of the line—about 200 feet—and towed the boat about half a mile further out to sea. He then came to the surface and we over-handed the line and pulled up close to him again. When he caught sight of the boat he turned and came toward us and threw one of his flippers over the gunwale of the boat, nearly capsizing her. I climbed up on the upper side and shoved him off with an oar. He grabbed the end of the oar and bit the end of it off like a piece of cheese. His movements in the water were very swift; using his fore flipper he could turn almost instantly from one side to the other and his head would project about 18 inches from the body. I succeeded in throwing a noose over his head and later by attracting his attention in the opposite direction got ropes around both flippers—finally having five lines on him—and started to tow him toward the shore. He repeatedly slipped the ropes off from his neck and flippers—several times getting almost entirely free. We were from 11:30 A. M. till nearly 4 P. M. in finally landing him. When about half way to shore he suddenly turned and made a break out to sea, towing the boat stern first with all sail drawing full for several hundred yards with little effort. He emitted at intervals a noise resembling the grunt of a wild boar. There were (when we first tackled him) about a dozen ramoras attached to different parts of the body. Most of them stayed with him all through the struggle and only deserted him when I hoisted him to the deck of the dock. I captured two of them and kept them in a bucket for several days. One was about ten inches long. The turtle lived for four days after taking out of the water—being very lively when first landed and gradually subsiding. I don't think this species ever come out of the water on their own responsibility

"So far as I can learn there has been but one other of this kind ever taken on this coast. It was less than half the size of this and was entangled in a fisherman's net and was wounded in capturing, so that it died soon after. The meat was sold to the hotels here and was very fine eating.

"Respectfully,

"G. W. GOURLEY."

Inquiry regarding the second specimen referred to in Mr. Gourley's letter finally resulted, through the kindness of Dr. Frank M. Anderson, in my securing from Mr. E. B. Hoyt of San Luis Obispo, a photograph of this turtle, taken soon after its death. Mr. Hoyt tells me that this photograph was taken by himself at Santa Barbara in July or August, 1901. It shows the animal covering more than half the length of the floor of a dray on which it was lying. This photograph is reproduced in Plate X.

The third individual of this species was preserved in the museum at Coronado, San Diego County, which I am told is now a thing of the past. All that I have been able to learn of its history is contained in the following note from Mrs. E. S. Newcomb, who was in charge of the collection:

"CORONADO, March 21, 1896.

"*Dear Sir:*—I am only posted in regard to one marine turtle, which hangs in the entrance of our museum, and provokes various witty remarks from the travelling public.... This turtle was caught off Point Loma [San Diego Co.] by a fisherman, weight 800 lbs. He sold it to the market, where Prof. Ward recognized the skin as belonging to the Harp or Lute turtle, and purchased it for this museum. It has been here eight years. I am sorry my information is so meagre, but it is the best I can give you.

"Yours sincerely,

"(MRS.) E. S. NEWCOMB."

With no material for comparison I am unable to form an opinion as to the identity or specific distinctness of the Leather-back Turtles of the Atlantic, Indian, and Pacific oceans; but Distant's photograph of an individual from

South Africa certainly shows a style of coloration very different from that seen in those reproduced here.

A view of the superior surface of the hyoid is given (Plate XI) which makes it evident that the specimen figured by Gervais was incomplete.

SAN FRANCISCO, CALIFORNIA,

August 4, 1905.

# EXPLANATION OF PLATE IX.

Photograph of Leather-back Turtle captured at Santa Barbara, California, January, 1905.


PROC. CAL. ACAD. SCI. 3ᴅ. SER. ZOOL. VOL. IV.

[VAN DENBURGH] PLATE IX.

## EXPLANATION OF PLATE X.

Photograph of Leather-back Turtle captured at Santa Barbara, California, in July or August, 1901.

PROC. CAL. ACAD. SCI. 3D. SER. ZOOL. VOL. IV.

[VAN DENBURGH] PLATE X.

# EXPLANATION OF PLATE XI.

Hyoid of Leather-back Turtle captured at Santa Barbara, California, January, 1905.

Proc. Cal. Acad. Sci. 3ᴅ. Ser. Zool. Vol. IV.

[Van Denburgh] Plate XI.

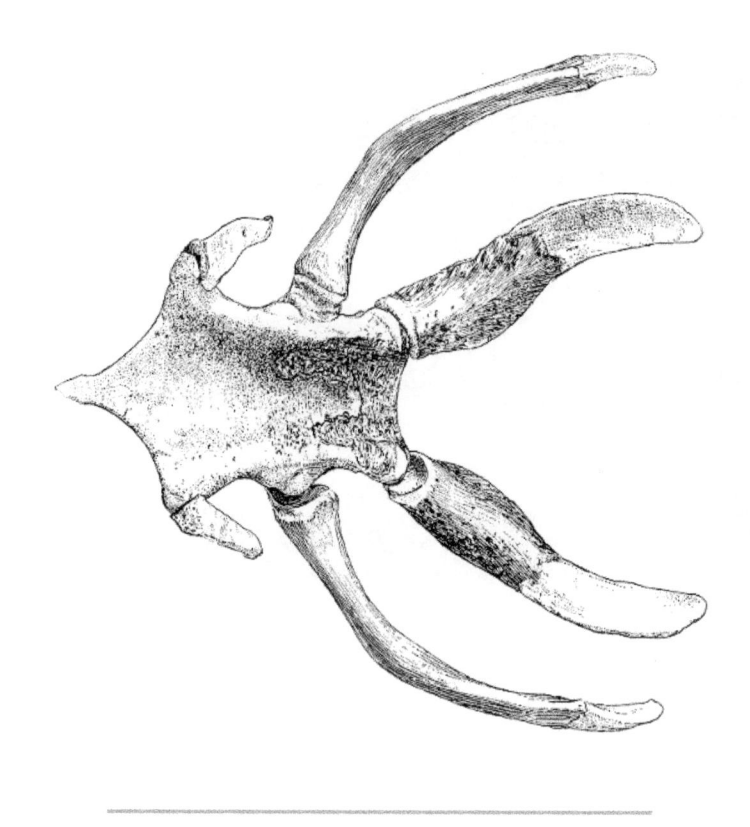

# PROCEEDINGS OF THE CALIFORNIA ACADEMY OF SCIENCES
## THIRD SERIES

Zoology                                  Vol. IV, No. 4

*Issued March 14, 1906*

# DESCRIPTION OF A NEW SPECIES OF THE GENUS PLETHODON (PLETHODON VANDYKEI) FROM MOUNT RAINIER, WASHINGTON

BY JOHN VAN DENBURGH

*Curator of the Department of Herpetology.*

In a small collection of amphibians secured in Washington by Dr. Edwin Cooper Van Dyke, Curator of the Department of Entomology, is an apparently undescribed species of salamander, which I take pleasure in naming, in honor of its collector,

**Plethodon vandykei** sp. nov.

*Diagnosis.*—Similar in general appearance to *Plethodon intermedius*, but much larger and stouter; costal grooves 12-13; toes and fingers webbed, only 2 phalanges of third and fourth toes free; adpressed limbs separated by 1 costal interspace; tail but slightly compressed; paratoid well developed; a dorsal band, not red; lower surfaces black.

*Type.*—Cal. Acad. Sci. No. 6910, Paradise Valley, Mt. Rainier Park, Washington, Dr. E. C. Van Dyke, July 15-31, 1905.

*Description.*—General form similar to *P. oregonensis*, but body not quite so much flattened, tail less compressed, and limbs shorter and stouter; tail cylindro-conic, somewhat compressed in posterior half, nearly equal to length of head and body; head depressed, about width of widest part of body; snout broadly truncate from above, rounded in profile; eyes moderate, smaller than in *P. oregonensis*, rather prominent, separated anteriorly by nearly twice the length of the orbital slit; nostrils small, near corners of snout, separated by about their distance from pupil; subnasal groove descending nearly to margin of lip; line of lip descending slightly below corner of snout and ascending below posterior edge of orbit; palatine *teeth* in 2 slightly curved series beginning some distance behind and a little internal to the internal nares, converging obliquely backward, and scarcely separated on the median line; parasphenoid teeth in 1 patch throughout, separated from palatine teeth by an interval equal to distance from nostril to edge of lip; internal nares rather small; tongue large, ovate, not emarginate, attached along median line but free laterally and for a short distance behind; neck a little narrower than body, with large elongate parotoid gland divided by a longitudinal groove running posteriorly and downward from eye to gular fold, other grooves behind, above and in front of parotoid; a groove along vertebral line; *costal grooves* between limbs 12 on right, 13 on left, not continued to midline either above or below; limbs a little shorter and stouter than in *P. oregonensis*, anterior with 4 and posterior with 5 digits; digits rather short, with broad rounded ends each with a terminal pad below, inner shortest, third longest, second finger longer than fourth, second toe shorter than fourth which is but little shorter than third; web well developed, extending nearly to end of inner digits, 2 phalanges of third and fourth toes free, feet very broadly palmate; tail slender, slightly compressed in posterior two-thirds, with rather indefinite grooves on proximal half; skin shiny, but roughened above and laterally and pitted below by the mouths of small glands; adpressed limbs separated by about the distance between 2 costal grooves.

A broad band extends along the whole dorsal surface from the snout to the tip of the tail. In the alcoholic specimen this band is dark clay-color, dotted with black on the upper surface of the head. It is broadest on the back of the head and narrowest above the anus. The upper surfaces of the limbs and the side of the snout are clay-color dotted with black. A black line runs from the eye to the nostril. The hands and feet are black dotted with clay-color. The chin and central gular region are white with a few scattered black dots. The sides of the neck and the sides and lower surfaces of the body and tail are intense black with a few scattered whitish dots on the belly and sides of tail and with a zone of crowded white dots along the sides of the neck and body.

Snout to anus                          60

| | |
|---|---|
| Front of anus to end of tail | 56 |
| Width of head | 9 |
| Nostril to orbit | 2 |
| Snout to orbit | 4 |
| Snout to gular fold | 13 |
| Snout to fore limb | 17 |
| Gular fold to anus | 47 |
| Axilla to groin | 34 |
| Adpressed limbs separated by | 3 |
| Fore limb | 15½ |
| Hind limb | 18½ |
| Heel to end of longest toe | 7 |
| Breadth of foot | 6 |

SAN FRANCISCO, CALIFORNIA,
December 21, 1905.

# PROCEEDINGS
## OF THE
## CALIFORNIA ACADEMY OF SCIENCES
### THIRD SERIES

Zoology                              Vol. IV, No. 5

*Issued March 14, 1906*

# ON THE OCCURRENCE OF THE SPOTTED NIGHT SNAKE, HYPSIGLENA OCHRORHYNCHUS, IN CENTRAL CALIFORNIA; AND ON THE SHAPE OF THE PUPIL IN THE REPTILIAN GENUS ARIZONA

### BY JOHN VAN DENBURGH

*Curator of the Department of Herpetology.*

### ON THE OCCURRENCE OF THE SPOTTED NIGHT SNAKE, HYPSIGLENA OCHRORHYNCHUS, IN CENTRAL CALIFORNIA

The little snake to which Cope, in 1860, gave the name *Hypsiglena ochrorhynchus* was first described from specimens secured at Cape San Lucas, Lower California. It has since been found to range across Arizona and northern Mexico to Texas. As recently as 1893, so little was known of the distribution of this snake in California that Dr. Stejneger, in recording the single specimen secured by the Death Valley Expedition in the Argus Range, Inyo County, California, thought that it added a species to the known fauna of the State. This snake had, however, already been taken at San Diego, California, as mentioned by Professor Cope in 1883. More recently, the species has been recorded by Cope from Witch Creek, San Diego County, and by myself from the Cuyamaca Mountains, San Diego County; Strawberry Valley and San Jacinto, Riverside County, and Hesperia, San Bernardino County.

These localities are all in the Desert and San Diegan faunal areas. It was with much interest, therefore, that I found this snake in the Californian Fauna close to the edge of the Pacific Fauna. The specimen was secured near Los Gatos, Santa Clara County, several hundred miles beyond the range of the species as previously known. It was found under a pile of recently cut hay, at an altitude of about eight or nine hundred feet, in what is locally known as the warm belt of the foothills, where *Bascanion laterale*, *Cnemidophorus tigris undulates*, and *Amphispiza belli* also occur.

## On the Shape of the Pupil in the Reptilian Genus Arizona

There has been, among herpetologists, much diversity of opinion as to the merits of Kennicott's genus *Arizona*. The validity of the single species for which he proposed the name *Arizona elegans* has, I believe, never been questioned, but the known generic characters have been rather inadequate. Accordingly, while some authors have followed Kennicott, others have referred the species variously to the genera *Pituophis* of Holbrook, *Rhinechis* of Michahelles, or *Coluber* of Linnæus.

I believe that all authors (myself included) who mention the point at all describe the eye of this snake as showing a round pupil. This is true of most alcoholic specimens, for in these the pupil usually is dilated. In two living specimens, however, I find that the pupil is slightly irregular in outline so that it appears somewhat eccentric, that it varies considerably in size from time to time, and that it is distinctly elliptic, with the long diameter vertical, but becomes nearly round when dilated. Some alcoholic specimens also show the pupil somewhat contracted and elliptic.

This point is of some importance, since the possession of a vertically elongate pupil is in itself ample basis for the recognition of the genus *Arizona* as distinct from the other colubrine genera with which it has been confused.

San Francisco, California,

February 24, 1906.

Type.

In fifty specimens the costal grooves are 17 in forty, 16 in six, and 18 in four.

Type.

Dr. Merriam tells me that a parallel is found in the island foxes, whose characters are constant on San Miguel but not on the other islands.

Skilton's description, which seems to apply rather to the species afterward named by Baird and Girard *Gerrhonotus principis*, is as follows:

"**Tropidolepis scincicauda**, n. s. Slender, tail much longer than body, cylindrical. Dermal plates of the body and tail, carinate above, smooth beneath, verticillate. The carinate plates in nine rows. Color, dusky green above, light ash color below. A row of small dark spots on each flank. Another row of smaller ones along the vertebral line. Some of the dark colored scales on the flanks tipped with a whitish color. Length five to five and a half inches."

The plate accompanying Skilton's article is so poor as to throw no light on this question, and it seems best to make no change in the nomenclature until some one has examined Skilton's specimens, one of which, according to Yarrow's Catalogue, is No. 3089 of the National Museum collection.

See Report, Chief of Engineers, U. S. A. 1876, pt. 3, pp. 435, 445, etc.

Since this was written I have been informed by Dr. F. Baker, of San Diego, that he has taken the following reptiles on these islands:—
North Coronado:
*Gerrhonotus scincicauda* [*ignavus?*], July 3, 1898,
*Eumeces skiltonianus*, July 3, 1898.
South Coronado:
*Uta stansburiana*, July 3, 1898,
*Gerrhonotus scincicauda* [*ignavus?*], July 3, 1898,
*Cnemidophorus stejnegeri*, July 3, 1898,
*Hypsiglena ochrorhynchus*, August 13, 1898,
*Crotalus* [*oregonus*], August 13, 1898.

Type.

Reproduced.

[9a] Reproduced.

Ann. & Mag. Nat. Hist. 2d ser. v. 10, 1852, p. 440.

Miss. Sci. au Mex. Recherches zool. 3d pt. p. 460.

Cat. Liz. Brit. Mus. v. 2, 1885, p. 299.

Report, U. S. Nat. Mus. 1898 (1900), p. 674.

Occas. Papers, Cal. Acad. Sci. 5, 1897, p. 116.

Abh. Nat. Verein Hamburg, v. 9, Hft. 1, 1885, p. 9.

Occas. Papers, Cal. Acad. Sci. 5, 1897, pp. 116, 118.

Ann. & Mag. Nat. Hist. 5th ser. v. 20, 1887, p. 50.

Report, U. S. Nat. Mus. 1898 (1900), p. 675.

Fauna Japonica, 1833, pp. 9, 12.

Proc. Zool. Soc. Lond. 1870, p. 410.

Sketch Nat. Hist. Australia, p. 188.

Austral. Vertebr. p. 39.

Prodrom. Zool. Victoria, v. 2, 1885, p. 2.

Cheeseman, Trans. New Zealand Inst. v. 25, 1893, p. 108.

Cat. Chelon. Brit. Mus. 1889, p. 10.

Essai sur l'Hist. Nat. du Chili, 1789, p. 194.

Ann. Univ. Chile, v. 104, 1899, [separate pp. 3-6], pl.—.

West Am. Scientist, v. 3, no. 24, 1887, p. 99.

It is interesting to note the similarity of the account given by Captain Subritzky of the capture of his specimen, which is given by Cheeseman (Trans. New Zealand Inst. v. 25, 1893, p. 109) as follows: "When passing Cape Brett on a voyage from Awanui to Auckland, he noticed a floating object, which he at first took for a boat bottom upwards. The schooner's boat was lowered, and he proceeded to inspect it; when, to his astonishment, it suddenly disappeared, shortly afterward reappearing a little distance further away. Returning to his vessel, he secured a harpoon and line, and then pulled cautiously up to the creature, soon recognizing it to be a large turtle-like animal entirely new to him. After a little manœuvering he succeeded in harpooning it in the neck. According to him, it made a most determined resistance, making for the boat open-mouthed, snapping its jaws violently. It succeeded in getting its flappers over the side of the boat, nearly capsizing it, but was stunned by a blow on the head, towed alongside the schooner, and hoisted on board."

Distant, Zoologist, 4th ser. v. 2, 1898, p. 500.

Gervais, N. Arch. Mus. v. 8, 1872, pl. VII, fig. 2.

NOTE.—Only a few copies of the original edition of this paper (Third Series, Vol. IV, Nos. 4 and 5, Zoology, pp. 61-67) had been distributed prior to the great fire of April 18, 1906, in which practically the entire edition was lost. To enable libraries and individuals to complete their files of the Proceedings this *exact reprint* is issued March 26, 1915.

BARTON W. EVERMANN, *Editor.*

Proc. Acad. Nat. Sci. Phila. 1860, p. 246.

N. A. Fauna, no. 7, 1893, p. 204.

Proc. Acad. Nat. Sci. Phila. 1883, p. 32.

Report, U. S. Nat. Mus. 1898 (1900), p. 954.

Occas. Papers, Cal. Acad. Sci. 5, 1897, p. 180.

Milton Keynes UK
Ingram Content Group UK Ltd.
UKHW040816051024
449151UK00004B/263